SEX ON THE MOON

ALSO BY BEN MEZRICH

The Accidental Billionaires

Rigged

Busting Vegas

Ugly Americans

Bringing Down the House

The Carrier

Skeptic

Fertile Ground

Skin (an *X-Files* book)

Reaper

Threshold

SEX ON THE MOON

THE AMAZING STORY BEHIND THE MOST AUDACIOUS HEIST IN HISTORY

BEN MEZRICH

DOUBLEDAY
NEW YORK LONDON TORONTO SYDNEY AUCKLAND

 Published in the United States by Doubleday, a division of Random House, Inc., New York, and in Canada by Random House of Canada Limited, Toronto.

www.doubleday.com

DOUBLEDAY and the portrayal of an anchor with a dolphin are registered trademarks of Random House, Inc.

Grateful acknowledgment is made to Thad Roberts for permission to reprint his personal letters.

Jacket design by Michael J. Windsor

LIBRARY OF CONGRESS CATALOGING-IN-PUBLICATION DATA
Mezrich, Ben.
Sex on the Moon : the amazing story behind the most audacious heist in history / Ben Mezrich. — 1st ed.
p. cm.
1. Roberts, Thad, 1977– 2. Theft—United States—Case studies. 3. Theft—Texas—Houston—History—21st century. 4. Lunar petrology—History—21st century 5. Lyndon B. Johnson Space Center—History—21st century. 6. Thieves—Texas—Houston—Biography. 7. Scientists—United States—Biography. 8. Interns—United States—Biography. 9. United States. National Aeronautics and Space Administration—Officials and employees—Biography. I. Title.
HV6248.R585M49 2011
364.16'28552—dc22
2010052240

ISBN 978-0-385-53392-8

PRINTED IN THE UNITED STATES OF AMERICA

10 9 8 7 6 5 4 3 2 1

First Edition

To Asher—this one will always be special, because you came into our world somewhere between Chapter 1 and Chapter 10. And maybe, just maybe, by the time you're old enough to read this, together we'll be watching someone take those first steps on Mars . . .

AUTHOR'S NOTE

Sex on the Moon is a dramatic, narrative account based on multiple interviews, numerous sources, and thousands of pages of court documents. I have tried to keep the chronology and the details of this narrative as close to exact as possible. Thad Roberts was generous with his time in helping me reconstruct this amazing story; thus many of the inferences are from his perspective, and I have done my best to describe these events as true to his individual perceptions as I could, without endorsing them myself. Since this is, at its heart, the story of Thad's journey, much of it is from his point of view. I am especially grateful for his permission to quote letters he wrote from prison; they are interspersed throughout the book.

In some instances, details of settings and descriptions have been changed to protect identities; certain names, individuals' characterizations, physical descriptions, and histories have been altered to protect privacy, in some cases at the character's own request. I do employ the technique of re-created dialogue: I have based this dialogue on the recollections of the participants I interviewed, but many of these conversations took place ten years ago, and thus some were re-created and compressed.

I address sources more fully in my acknowledgments, but it is appropriate here to again thank Thad Roberts for his incredible generosity. I would also like to thank Axel Emmermann, Gordon McWhorter, and Matt Emmi for their time, as well as numerous other sources who have asked to remain anonymous.

SEX ON THE MOON

PROLOGUE

It had to be the strangest getaway in history.

Thad Roberts tried to control his nerves as he stared up through the windshield of the idling four-wheel-drive Jeep. The rain was coming down in violent gray sheets, so fierce and thick he could barely make out the bright red traffic light hanging just a few feet in front of him. He had been sitting there for what seemed like forever; a long stretch of pavement serpentined into the gray mist behind him, winding back past a half-dozen other traffic lights—all of which he'd had to wait through, in exactly the same fashion. Even worse, between the lights he'd had to keep the Jeep at an agonizing five miles per hour—a veritable crawl along the desolate, rain-swept streets of the tightly controlled compound. It was unbelievably hard to drive at five miles per hour, especially when your neurons were going off like fireworks and your heart felt like it was going to blow right through your rib cage. But five miles per hour was the mandatory speed limit of the compound—posted every few yards on signs by the road—and at five miles per hour, once you hit one red light, you were going to hit them all.

Thad's fingers whitened against the Jeep's steering wheel as he watched the red glow, willing it to change to green. He wanted nothing more than to gun the engine, put his foot right through the floor, break the speed limit, and get the hell out of there. But he knew that there were cameras everywhere—that the entire getaway was being filmed and broadcast on more than a dozen security consoles. For this to work, he had to stay calm, obey the rules. He had to appear as if he belonged.

He took a deep breath, let the red glow from the traffic light splash across his cheeks. Only a few more seconds. He used the opportunity to toss a quick glance toward the passenger seat—which didn't help at all. Sandra looked even more terrified than he felt. Her face was ivory white, her eyes like saucers. He wanted to say something to calm her down, but he couldn't think of the words. She was pretty, with blondish-brown hair; even younger than Thad, barely nineteen years old. Maybe not the ideal accomplice for something like this—but she was an electronics specialist, and she had practically begged to be a part of the scheme.

Thad shifted his eyes toward the center "seat" between them, and almost smiled at the sight of his girlfriend crouched down beneath the dashboard, her lithe body curled up into a tight little ball. Rebecca had jet-black hair, cut short against her alabaster skin, and she was even prettier than Sandra. She had just turned twenty. But as young as she was, she was the only one of the three of them who didn't look scared. Her blue eyes were positively glowing with excitement. To her, this was beyond thrilling—really, James Bond kind of shit. Looking at her, Thad was infused with adrenaline. They were so damn close.

And suddenly he was bathed in green as the light finally changed. Thad touched the gas pedal, and the Jeep jerked forward—then he quickly lifted his foot—making sure the speedometer read exactly 5 mph. The slow-motion getaway continued, the only sounds the rumble of the Jeep's engine and the crackle of the rain against the windshield.

A bare few minutes later, they came to the last traffic light—and again, of course, it was red. Even worse, Thad quickly made out the security kiosk just a few yards to the left of the light. He could see at least two uniformed guards inside. Thad held his breath as he slowed the Jeep to a stop at the light; he kept his head facing forward, willing Sandra to do the same. He didn't want to have to explain why he was at the compound, past midnight on a Saturday. Thad was counting on the fact that neither of the guards would be eager to step out into the rain to interrogate him. Even so, if one of the guards had looked carefully, he might have noticed that the Jeep was sagging in the back. In fact, the vehicle's rear axle was bent so low that the chassis almost scraped the ground as they idled at the traffic stop.

The sag of the Jeep was one of the few things that Thad and his two accomplices hadn't planned. A miscalculation, actually: the safe that Thad and the two girls had hoisted into the back of the Jeep—less than ten minutes ago—weighed much more than Thad had expected, probably close to six hundred pounds. It had taken all three of them and a levered dolly to perform the feat, and even so Thad had strained every muscle in his back and legs getting the damn thing situated properly. Thad was just thankful that the Jeep's axle hadn't collapsed under the weight. As it was, he was pretty sure that even a cursory inspection of the vehicle would be enough to blow the whole operation.

Thankfully, neither of the guards made any move to step out of the kiosk. When the light shifted to green, Thad had to use all of his self-control to barely touch the gas—piloting them forward at the prescribed 5 mph. Almost instantly, the exit gate came into view. They approached, inch by inch—and at the last minute, the gate swung upward, out of the way. And then they were through. Thad slowly accelerated. Ten mph.

Twenty mph.

Thirty mph.

He glanced in the rearview mirror. The compound had receded into the rain.

He looked at Sandra—and she stared back at him. Rebecca uncurled herself and sat up in the middle of the Jeep, throwing an arm over his shoulder. Then they were all screaming in joy. They had done it. My God, they had truly pulled it off.

When the celebration had died down, Thad glanced into the rear-view mirror again—but this time, he wasn't looking at the road behind them. He could see the dark bulk of the safe, covered in a plastic tarp they had bought in a hardware store just twenty-four hours ago. The sight of the thing caused his chest to tighten—a mix of anticipation and what could only be described as pure awe.

In that safe was the most precious substance on Earth. A national treasure—of unimaginable value, something that had never been stolen before—something that could never, in fact, be replaced. Thad wasn't sure what the contents of the safe were worth—but he did know that if he'd wanted to, he could have just as easily walked off with enough of the stuff to make him the richest man in the world. As it was, he and his accomplices had pulled off one of the biggest heists in U.S. history.

But to Thad, it hadn't really been about the monetary value of the contents of the safe. All he'd really wanted to do was keep a promise to the girl sitting next to him, her arm over his shoulder. A simple promise that millions of other men had made to millions of women over the years.

He had promised to give her the moon.

The difference was, Thad Roberts was the first man who was actually going to keep that promise.

I may never hold you again, my love, I may never again feel the warmth of your touch, the softness of your voice, the adventure in your eyes, but they will always be a part of me. Eternity lives in every true connection, every moment that opens your eyes to something new and deepens your internal spring. My very being soared beyond the horizon with you, Rebecca. Everything that I am will always carry that echo. I cannot abandon that. I cannot cover my heart. I will always love you. I will always remember you.

1

Five Years Earlier, February 1997

There was something vaguely menacing about the folders. Off-white and three-ringed, row upon row, rising up the skyscraper-like corrugated-metal shelving units that obscured all four walls of the cramped, nearly windowless first-floor room. It wasn't the folders' color that was the problem, exactly; a shade that couldn't be found in nature, even in a place as abundantly natural as Salt Lake City, Utah. Nor was it the black block lettering that ran up the spine of each folder, declaring the contents in language a third grader could understand. It was the idea behind the manila metropolis itself. What the folders represented: a literal way station on the search for the meaning of life.

Maybe not the meaning of life—but certainly its direction. Thad Roberts stood in front of one of the towering shelving units, his hands nervously jammed into the deep pockets of his green, oversized Windbreaker. His windswept, free-form mop of light brown hair cast tangled shadows down across his high cheekbones. He supposed that such a room existed in cities all over the country—maybe all over the world. Probably every university campus had a place like this. No doubt, many were more glamorous than the rectangular, folder-filled box that was the career center of the University of Utah, but the essence of the place was quite probably duplicated all over the globe. A mildly terrifying place where lost souls gathered to seek a future, or at least the sort of future that could be summed up between the covers of a shiny three-ring folder.

It was barely ten minutes past seven in the evening, but Thad was already swaying in his mud-scuffed Timberland boots as he surveyed the shelves, for what had to be the hundredth time. He had been in the career center two hours already, and by now he was approaching the damn folders almost at random. He'd pulled a half dozen of the folders off the shelves, piling them on one of the small wooden desks that lined the interior of the room behind him: Financial Adviser, Geologist, Air Traffic Controller, Physical Therapist. None of the choices sang to him, and he was truly close to the breaking point. He was fighting the urge to start sweeping the rest of the folders off the shelves with both hands. Close his eyes, make do with whatever landed on top of the pile.

Roll the dice, get a life.

He blinked, hard, trying to push the bleariness out of his normally brightly lit, citrine-green eyes. Or maybe it was time to just give up. He'd been at this way too long. And he wasn't any closer to figuring out what he was going to do with himself.

At twenty, he was drowning in student debt, even though he hadn't even fully graduated from the university, leaving early to take on multiple jobs just to survive. That day, he'd been up since four A.M., spending most of the past fifteen hours running around a backwoods construction site, basically a glorified gofer. He had about three hundred dollars in his bank account: the Windbreaker and boots he was wearing were three years old, and the shirt beneath his Windbreaker was held together by multiple assaults with a needle and thread, courtesy of Sonya, his beautiful but equally broke wife. He had no money, and certainly no safety net: he hadn't spoken to his parents in more than a year, and he was pretty sure he wasn't going to speak to them ever again. In fact, by their own admission, Thad didn't really have parents anymore.

Instead, what he had was in front of him, a skyscraper-high bookshelf lined with three-ring folders.

He wasn't even sure what he was looking for. He'd always been a stellar student, acing courses in everything from business to philosophy. While he was growing up, everyone had always told him how smart he was, and though some bad breaks had derailed him, he knew that he had the capability to learn. Wasn't that supposed to be the most important thing?

He pushed the hair away from in front of his eyes and turned back toward the very first row of folders. As tired as he felt, he decided he would start over and go through them all again.

To his surprise, almost immediately one of the labels caught his eye, about five folders in from the beginning of the shelf. It was a folder Thad had paused at when he'd first walked into the career center, but he hadn't yet pulled it out. He'd discounted it before, because he'd thought it was ridiculous, and probably way out of his reach. But now, a couple of hours later, his inhibitions were dwindling.

He reached for the folder and reread the block letters.

ASTRONAUT.

That there was even a job folder for such a career seemed improbable. Thad had initially skipped over it because he was pretty sure you had to be in the air force to even consider being an astronaut—but at this point, he figured it couldn't hurt to look. After all, he did love the sky. One of the first things he'd done when he'd arrived at the U of U was to visit the school's observatory, and he'd dropped by the small hilltop facility a few times since, usually when he needed space to think. Literally.

He began to leaf through the folder. To his surprise, it was divided into two parts: Pilots and Mission Specialists. The pilots were almost exclusively military, because they were the ones who flew the equipment. But the mission specialists could come from a variety of fields. These were the people who got their feet dirty, who went out into the different environments and conducted experiments. Thad figured that during moon landings, the two guys who walked around hitting golf balls were mission specialists. The guy who stayed behind in the spacecraft was the pilot. Thad wondered how jealous that would make you, going all the way to the moon but never getting to step outside. If Thad were an astronaut, he wanted to be the guy who walked on the moon.

As he read deeper into the file, he felt his mind snapping into focus. He realized right away that if he really was going to do this—and it was a crazy thought, but still—he'd have to go back to school. He'd have to get a degree in something that NASA would be interested in. Biology, astrophysics, maybe geology. He would also have to gain expertise in a variety of other endeavors. Scuba, because the astronauts trained underwater. Languages, because space was international now, and there would be plenty of exchanges of people and machinery. A pilot's license—even though he wouldn't be able to compete with the military kids, he'd need to know how to fly.

It all seemed so fascinating, so romantic. Growing up, he'd never really dreamed about the stars—he was too young to remember anything significant about the first steps on the moon. But he was instantly engaged by the idea because it seemed to fit him in so many ways. He was a dreamer, but he knew how to get dirty. He wanted to learn all these things—scuba, flying, Russian—and here was a reason to do it all.

Shit, who wouldn't want to walk on the moon?

Of course, there was very little about the moon in the folder. The few articles about NASA's current state seemed much more organized around another destination altogether: Mars. NASA scientists were hoping to one day launch an effort similar to the '69 moon landing to try to get to Mars. Thad wondered what it would be like to be an astronaut on that mission. To have a chance to be the first person someplace new, someplace untouched. Someplace far away from Utah.

To be the first man on Mars.

Thad suddenly realized he wasn't nervous anymore.

. . .

Legs furiously pumping, the pedals a near blur beneath his feet, his body leaning all the way forward over the handlebars, the frigid air tearing at his bare cheeks and forehead, Thad was moving so fast he could barely see the pavement flashing beneath him. He kept his eyes focused on the cone of orange light spitting out of the little headlamp attached to the front of the bike, ignoring the trees flashing by on either side, the flicker from windows hidden deep between the leaves. He took the last hill at top speed, the rubber tires skidding briefly against the iced road, and then the orange cone flashed against gravel—the driveway that led up to his rented single-bedroom home. He hit the brakes a second too late, but he was still able to take the gravel, his back tire jerking side to side. A moment later, he was clear of the bike, his boots hitting the grass of his front lawn.

The house was little more than a shack, but Sonya was waiting on the front porch, her beautiful reddish-blond hair pulled back in a ponytail and her white sweater tight against her curves. Thad ran up to her and held his hands out. She grinned, pulling the bottom of her sweater up to reveal the flat plane of her toned stomach. Then she took his hands and pressed them against her warm skin, shivering as she did so. It was a cute little ritual they had developed over the past few months of living together. Maybe it was stupid and maybe it was sweet, but Thad was certain he'd remember these moments for the rest of his life.

A minute after that and they were inside. The living room was pretty bare: a few pieces of wooden furniture they had picked up at yard sales, a TV that was almost never on, a freestanding radiator that spat arcs of hissing water when it was turned too high. Thad led his wife to the couch by the TV and, sitting beside her, told her that he wanted to be an astronaut. He explained in detail what that meant, the things he would need to do and what they would have to reconfigure to make those things possible. It was going to take sacrifice, on both their parts. Sonya already had a full-time job as a dental assistant, and she had just started modeling in the evenings, had even signed with a local agency. But this would mean he would have to start school again, and take scuba and flying lessons. He would have to fill his résumé with the things that would impress scientists at NASA. It wasn't going to be easy.

"You want to be an astronaut," Sonya repeated, looking at him.

He half expected her to burst out laughing. Instead, she ran a hand through his tangled hair.

"Cool. I guess I'm going to need to get another job."

2

One year earlier, astronauts, Mars, and NASA scientists had been the furthest things from Thad's thoughts as he huddled, trembling, in the backseat of his parents' oversized gray van, waiting for his father to murder him.

The van was parked in the driveway in front of Thad's family's house, a ranch-style building on the outskirts of Syracuse, Utah. Syracuse was an isolated speck of a place, nearly impossible to find on a map, a pseudo farm town—which meant that everybody there was a pseudo farmer, except for the few families that *actually* lived on farms. Thad's family lived on an acre-and-a-half garden, where they grew their own vegetables, next door to a small cow pasture that provided them with just enough meat to feed Thad and his six brothers and sisters. It was a simple existence, and on paper it might even have seemed pretty and quaint. Thad hadn't seen it that way in a very long time.

It had just started to snow outside the van's windows, an angry whirl of gargantuan white flakes. Thad barely noticed, because he was too busy staring at his house's front door. Any moment, he was certain his father was going to come through that door with a shotgun, march up to the van, and shoot Thad in the head.

Thad hadn't come to the conclusion that his father was about to murder him frivolously. In fact, he was nearly certain it was about to happen. He had watched the seething anger deepen in the redness that splotched across the back of his dad's neck the entire hour-ride home from the Salt Lake City airport. His mother, silent in the front seat next to his father, had glanced back only once during the ride, and her eyes had only confirmed the thought.

Thad believed he had finally pushed the man over the edge, and now his father was going to do what he believed was necessary.

Thad fought back tears as he stared through the thick snow, wondering if it would hurt, wondering if he'd even put up his hands or beg for forgiveness. In his opinion, his dad was a brutal man, but maybe what he was about to do was right. Maybe that was exactly what Thad deserved.

The truth was, in the back of his mind he had expected this moment since the day he had met Sonya his freshman year of high school. A shy, nerdy kid like him had had no business going after the pretty, popular redhead—but for some reason, she had fallen for him as well. In a normal part of the world, they would have been high school sweethearts, boyfriend and girlfriend—whatever label meant holding hands in homeroom and stealing kisses beneath the bleachers at football games. In the utterly Mormon enclave where Thad had grown up, things didn't work quite that way.

Thad's father had forbidden him to have a girlfriend; so Thad and Sonya had concocted a charade. For three years, Thad had pretended to go on dates with all of Sonya's friends so it would seem he had light and wholesome relationships. He had broken free from his shyness out of necessity—first as a ruse, and then for real. And without that shyness holding him back, he had given in to the impulse to do what felt natural—no matter how wrong his religion had told him it was.

The first time had been intense, tentative, explosive, and more than a little terrifying. In the back of Sonya's father's car, exposed skin sticking to the vinyl seats, condensation forming across the fogged-up back window, their bodies arching while their minds raced to stay ahead of their Mormon guilt.

From then on, Thad and Sonya had lived in fear. Thad believed that to his fiercely religious father, what he and Sonya had done—it was an explicit sin. Keeping the secret and the guilt pushed deeply away had been excruciatingly hard, but somehow Thad had made it to the day he had left for what was supposed to be his two-year mission—the rite of passage for every nineteen-year-old Mormon boy.

First, Thad had been sent to the MTC—the Mission Training Center—located in Provo, Utah. Dressed in the standard uniform—white button-down shirt, dark pants, sometimes a suit—Thad had found himself cut off from the rest of the world, relearning how to talk, to dress, to walk, to think—while sleeping eight teenagers to a room in a dormitory filled with bunk beds.

Almost immediately, Thad had begun to feel that he was unworthy—that the secret he was keeping was really a lie, to his family, to the church, and to God. But a little after two in the morning his third night at the MTC, he had been lying in his bunk staring at the ceiling, listening to the breathing of seven other teenagers away from their families for the first time in their lives—when suddenly a kid on the bunk directly across from him had broken the nightly silence.

"Guys, you awake? I've got something I gotta tell you, but you've got to promise you'll never tell anybody else . . ."

And with that, the kid had suddenly begun a confession. Just like Thad, this kid had had sex with his girlfriend before coming to the MTC. Technically, Thad and the other kids in the bunk beds were supposed to be shocked; instead, another kid began talking—and suddenly he, too, was making the same confession. He had had sex with his girlfriend as well.

By the end of that night, every kid in that room had confessed to having sex. And for the first night in years, Thad slept without guilt. The next morning, he began to wonder if the sin of premarital sex was really as unforgivable as he had thought. Maybe, like the kid in the MTC dorm, it was something he simply needed to confess.

Before he lost the nerve, he had decided to go through with it—

making an appointment with his mission president. Meeting in the man's stark office, Thad told the man about Sonya and the sin they had committed. He had truly believed he'd get sympathy at the very least, and a path to the penance he needed.

But the penance wasn't offered; instead, the president had immediately called together the church quorum necessary to kick Thad off his mission—effectively, branding him a sinner in front of the entire Mormon world. The very words the man used would reverberate in Thad's mind the rest of his life.

"You are no longer worthy to serve God."

He had been sent home the next day.

And now here he was: sitting in his parents' van, numb to the snow and the cold. Thad considered making a run for it, but then he'd never see Sonya again, and that seemed like something worse than the shame and the embarrassment—worse, even, than a bullet from his father. So he just sat there and waited. Five minutes became ten minutes became half an hour, and soon he lost track of how long he had been in the van. The snow began to pile up, blanketing the vegetable garden and the cow pasture and even the house, turning everything a brilliant shade of white. The air in the back of the van was becoming frigid, and Thad could see his own breath freezing into little starbursts of crystal on the windowpane—but still he sat there, his mind a jittery mess.

Not until the air outside started to dim, and the snow piled so thick against the van's windows that he could no longer see the house, did he decide that he had no choice but to follow his parents inside. Maybe his dad had decided that killing him out in the driveway was too public; this was something you had to take care of in the privacy of your own home.

Thad collected his single duffel bag—a couple more white shirts, some toiletries, a handful of copies of *The Book of Mormon*, and maybe a half-dozen ties—and exited the van. The snow stung his bare cheeks and neck, but he barely noticed. He crossed the front yard that led up to his house in a trancelike state.

He found his parents in the kitchen. His dad was sitting at the table, his mother next to him. Neither looked at him as he entered the room. Nobody spoke, and Thad stood for a moment just inside the doorway, listening to the melting snow drip against the porcelain-tiled floor. Then he let his duffel bag drop and took a seat across the table from his parents.

His dad glared at him, and the fury in the man's eyes was so nearly palpable it all but knocked Thad out of his chair. His chest was heaving, but he felt like he couldn't breathe, his stomach churned and the heat rose up his back in vicious twists that truly felt like flames. His mom was staring at her reflection in the glass table, refusing to meet his eyes. This wasn't about his mom, anyway. It was about Thad and his father, and what had to happen next.

"Because we're loving parents," Thad would remember his dad saying, through clenched teeth, "we are giving you two months."

Thad felt the air come back into his lungs. *Two months?* He wasn't even sure what that meant, but it wasn't the barrel of a shotgun. His father wasn't going to kill him, at least not today, and that felt like a good thing.

"Two months," his father repeated. "And these are the rules. You aren't allowed in your old room. You aren't allowed to have any of your old possessions. Just that duffel from your mission."

Thad nodded. So far it wasn't so bad. He was alive, and he was home. But his dad wasn't finished yet.

"You will sleep in the basement. You are not to talk to any of your brothers or sisters. You can't even look at them. No eye contact. No notes. No phone calls. No communication at all. Because you, Thad, are going to hell, and any communication you have with the rest of us will only make us go to hell, too."

Thad opened his mouth but couldn't find any words. It was a hard thing to hear, so explicit and out in the open. Hell, to his father, was not some arbitrary religious concept that you learned about in church; it was physically real, fiery and violent, and forever. And that was where Thad was headed.

"You will leave the house by six every morning," his father continued, his voice even and low. "You won't return until after ten at night. I don't care what you do during those hours, but you will not be here. No one will know you are still living in this home. No one will talk to you, or see you, or think about you. You simply do not exist."

Without another word, his father stood and turned his back on Thad. Thad's mother remained at the table, staring at the glass. Thad was in the room with them, but he was alone.

He didn't exist.

He picked up his duffel bag and headed to the door that led to the basement.

Later that evening, as he was about to take off his white shirt and climb onto the cot his father had left for him to sleep on for the next two months, he was surprised to hear footsteps on the stairs that led to the rest of the house. Even more surprising, the visitor was his mother, quietly coming down to the basement to see him.

For a brief moment, he felt that maybe everything was going to be okay—that she was coming to tell him that he was still part of the family, or possibly even give him a hug. He watched as she paused on the bottom step, looking at him. There were tears rolling down her face, and the hopefulness in him grew. She was going to give him a sign that she really did love him, that although they were treating him harshly, it was out of love.

And then a hardness came into her eyes, and she turned away as she spoke.

"When you die, are you going to blame how you turned out on me?"

With that, she headed back up the steps.

Thad stood there, watching her go.

. . .

Two months later, he officially moved out of the house and married Sonya. His parents were there to witness the vows, but they didn't stay for the cutting of the cake. They spoke barely two words to congratulate Sonya and her family, and then they were out the door, on their way back home to Syracuse. Thad was no longer their burden. It was going to be up to him to make a life for himself, whether that meant working as a gofer on a construction site—or something else entirely.

Something meaningful and important.

It was solely up to him.

3

There was nothing like a two-million-year-old rock to put things in perspective.

Thad grimaced as he took the last few steps across the dimly lit storage room, the oversized plastic crate balanced precariously in his outstretched arms. The crate was much heavier than it looked; it wasn't just one rock he was transporting through the bowels of the University of Utah Museum—the crate seemed like it was packed with a big enough collection to pave a short driveway. It was going to take hours to go through all the samples, entering the details into the computerized archive kept by the geology department—and there were two more boxes just like this one still waiting for him in the upstairs receiving closet. No doubt, he was going to be in the museum all night—which was exactly why he had volunteered for the inventory assignment. Anything to keep him from pacing the floors of his and Sonya's living room, waiting for the sun to rise.

He reached the shelving unit on the far side of the room and heaved the crate onto one of the corrugated shelves. His shoulders burned from the effort, but it was a good sort of pain; he knew he was contributing something, even if it was just a long night of physical labor. Like the anonymous people who had donated the samples in the three crates to the university museum, he was giving something of himself to the geology department; in return, whenever he walked through the brightly lit display corridors upstairs, he would feel a sense of pride.

Although, he realized, these particular rocks would never actually make it into the displays upstairs. When he'd arrived at the museum earlier that evening, he'd been told that the samples he'd be cataloging were donated materials deemed not good enough for the collections upstairs. Though some of the rocks seemed pretty interesting to Thad—a handful of fossils and semiprecious minerals that told stories of deep time, ancient life-forms, maybe even evolution itself—the museum thought of it as mostly junk. These rocks would probably remain in this crate in the bowels of the museum far into the foreseeable future.

But that didn't mean they wouldn't be inventoried, cataloged, and described in detail—as soon as the life returned to Thad's shoulders. It seemed a shame—these items hidden away in a basement—but it wasn't his decision to make. He was a volunteer, and no matter how pointless he thought it was to hide these donated fossils in a basement, he was glad to be the one getting his hands dirty for the greater good of the museum—in no small part because every minute he was in the basement, straining his muscles, was one less minute spent agonizing over the phone call that was now only hours away.

Thad felt a surge of adrenaline at the mere thought of the call, scheduled for eight A.M. He knew that if he was at home instead of in the museum basement, he really would have been burning off the soles of his shoes, circling the cordless phone on the desk in his living room. The day before, in preparation, he'd even pasted a pair of photos to the bare wall behind the desk. One showed a reasonably chiseled, crew-cutted man in his mid-thirties, smiling toward the camera, dressed in a conservative-looking suit and tie. The second photo was of a woman who appeared to be middle-aged; from the style of the picture and the discomfort in the woman's pose, it was obvious that the shot had been culled from a college administration handbook.

No doubt the photos were probably overkill, maybe even a little psychotic—but Thad wasn't going to take any chances, because the call really was that important. Disembodied voices made him nervous, so if he had to do the interview by phone, he was going to see the people he was talking to, even if it was in two dimensions.

Eight A.M.—crazy, that the call that could potentially change everything for him was now just a handful of hours away, because the truth was, he had spent the past two years preparing for this moment.

But that knowledge didn't make him any less anxious. It wasn't just any interview—and the position at the Johnson Space Center Cooperative Program wasn't just any job. It was the first step toward reaching his goal of becoming an astronaut. Since the sixties, the JSC co-op program had been supplying NASA with talent. It had grown into an incredibly competitive and prestigious feeder to the space game; on average, there were eight hundred applicants for every fifty spots in the program, and the majority of the applicants were engineering majors from the country's top universities. Co-ops got to spend at least three "tours" at the space center in Houston, working on projects that were directly related to the space program. Most of the co-ops went on to work at the space center, and a handful of standouts had ended up successfully entering the astronaut training program. Aside from the air force, which Thad had already ruled out, the JSC co-op program was his best—and really, only—avenue to becoming an astronaut.

No question, he had to ace the phone interview. And he had spent the past two years building himself into exactly the sort of person the JSC was looking for. Aside from a dizzying collection of physics, geology, and anthropology courses—he was majoring in all three disciplines—he'd filled his résumé with a wide array of outside accomplishments. He was the founder of the Utah Astronomical Society, and had personally built up the college's observatory into one of the premier science clubs on campus. He was routinely doing volunteer dinosaur digs with the paleontology group, an offshoot of the geology department. He had gotten his pilot's license and had become a certified expert in scuba diving. He'd taken Russian and Japanese. To top it all off, he'd recently completed a charity bike ride to raise money for cystic fibrosis; he and Sonya had biked all the way from the front door of the Salt Lake City hospital to San Francisco, bringing in just shy of $10,000 for the cause.

He had done everything he could to make himself the perfect candidate. Along the way, he'd fought down the gnawing sense that no matter what he did, he'd always be starting a few steps back from the other kids applying for the program; most would probably be coming from more elite schools, paid for by loving parents. Most wouldn't already be married at twenty-three. Hell, most wouldn't *be* twenty-three; they'd be college age, from middle-class backgrounds. Thad was different. He'd always be an outsider.

He'd have to work harder than everybody else to prove himself. Already, he'd shown them how persistent he could be.

He thought back to the photos attached to the wall above the desk in his living room. Bob Musgrove was the co-op program manager, responsible for all new hires. The woman in the photo next to Musgrove's was the man's secretary, who Thad assumed might be part of the phone interview as well. Thad had spoken to her often, and had heard Musgrove's voice on the man's voice-mail greeting more times than he could remember. Thad had lost count after leaving his hundredth voice-mail message—to go along with the hundreds of e-mails, dozens of letters, and even a handful of faxes to the JSC co-op fax line. None of the voice mails or e-mails had gotten him a response, but he had kept going, placing a call nearly every day.

And it had seemed to work out; four days ago, Thad had received a simple e-mail from Musgrove—as if Thad hadn't been trying to contact the man for months—telling him when to call in for his initial phone interview. A message from the man's secretary had confirmed the time and date, and now it was going to be up to Thad.

One simple phone call.

Thad took a deep breath—and the dust-filled, musty air of the museum basement brought him back into the moment. His heart still pounded, but thoughts of the phone call dissipated as he finally worked the stiffness out of his arms. Before heading back upstairs for the other two crates, he took a moment to peer over the top of the box he'd just settled onto the shelf. Sitting on top of the heavy pile of specimens was a jagged little piece of rock that had been given to the museum by an unnamed collector. Thad could barely make out the faint outline of some sort of fossil on the surface of the stone—maybe a prehistoric plant, maybe something better, like an insect or even a footprint. God only knew what it was—but that mystery made it even more amazing. It was a real piece of history, a step in evolution.

Yes, it sure as hell put things in perspective. Thad had been preparing for the upcoming phone interview for two years—and if he did well, if he kept his cool and said the right things, maybe he really was going to be on his way to becoming an astronaut.

That rock, on other hand, had survived two million years of erosion—to end up in a box in the dark basement of a museum.

Thad took another breath—and came to a sudden decision.

He glanced over his shoulder, making sure he was alone. Then he reached into the box, grabbed the fossil, and jammed it deep into his pocket.

. . .

Eight hours later, Thad's mind whirled as he leaned back at the desk in his living room—a stunned expression spreading across his face. Bob Musgrove's words still reverberated in his ears, as surprising now as they had seemed when they'd first echoed through the cordless phone now sitting dormant in front of him:

"Well, Thad, I think you'll be a great addition to the co-op program."

Just like that—after only the briefest of interviews. Musgrove hadn't asked him about his scientific background, about how he was going to compensate for the fact that he wasn't an engineer, about his moderately advanced age—all the man had wanted to talk about was the charity bike ride; how he and Sonya had raised money for cystic fibrosis while living out of a tent, collecting blisters on desolate roads crisscrossing the country. And then Musgrove had simply sprung it on him, out of nowhere.

"Your résumé is stellar. Your grades really picked up once you started taking courses you enjoyed, and it's obvious you know how to work hard. I'd already made the decision before I got you on the phone. You're exactly the kind of person we look for."

Thad couldn't believe it. All that anxiety, that built-up adrenaline—and now it was really going to happen.

"There are two types of people who work at NASA," Musgrove had finished cheerily. "People who are obsessed with space. And people who are about to become obsessed with space."

With that, the man had hung up, the phone going quiet in Thad's hand.

And that was it—Thad was on his way. He leaned back in his chair, grinning ear to ear. He was going to be a co-op at the Johnson Space Center.

Houston, we have liftoff . . .

4

The twelve-year-old kids in the *Star Trek* uniforms should have given it away. That, or the fact that the line Thad was standing in ended in a turnstile manned by a guy in a bright orange space suit. But Thad's anxiety level was so high, his mind whirling so fast, he didn't realize anything was wrong until the kids in the uniforms had disappeared into the building in front of him, and he was standing right up against the turnstile, staring past the orange suit into an atrium that looked way more like Disney's Epcot Center than a working government building. There was a mock-up of the Apollo lunar lander hanging from the ceiling, and something that resembled the interior of the space shuttle jutting right out of the far wall—as if the damn thing had crashed through from the other side, embedding itself for the amusement of the throngs of children scrambling over its fuselage. Stranger still, Thad noticed multiple vending machines hawking everything from colorful space ice cream to baseball hats with the NASA emblem emblazoned across the front. He'd either taken a wrong turn on his way out of the parking lot, or NASA wasn't the buttoned-down institution he had imagined after all.

He turned back toward the man in the orange space suit. On closer inspection, the guy couldn't have been more than nineteen years old.

"I think I might be in the wrong place."

"Depends where you're trying to go," the kid responded, barely looking at him. "You here for the zero-gravity show? Tickets are twenty bucks, but you have to get them at the ticket office."

Thad shook his head.

"I'm not here for the zero-gravity show. I'm here for work. I mean, I'm supposed to start today. I'm in the co-op program."

The kid in the space suit yawned.

"Uh, guy, this isn't the Johnson Space Center. This is Space Center Houston. The JSC is next door. But you have to be authorized to get through security."

"Shit, thanks."

Thad quickly stepped out of line and rushed back out toward the parking lot. Christ, now he was going to be late—and on his first day. He pushed through the glass double doors and winced as the morning heat hit him full in the face; even though it was the first week of September, it still felt like an oven outside. The sky was blindingly bright and it had to be over ninety degrees. Thad pulled a pair of sunglasses out of his shirt pocket. The shirt was white, with short sleeves, and his pants were khaki and a little too long, hanging down over his black dress shoes. He knew the shoes were entirely wrong, but they were the only pair he owned that weren't caked in dried mud from multiple dinosaur digs and geological fieldwork. Dress shoes would have to do.

He quickly found his car—a 1996 bright green Toyota Tercel with Utah plates—and navigated his way around the tour buses that cluttered the vast parking lot. In retrospect, he should've known he was in the wrong place by virtue of how easy it had been to drive right up to the squat, rectangular building; after all, this was NASA, it should have been one of the most secure complexes in the country. But Thad was working on barely any sleep, having spent half the night driving the last leg of the fifteen-hundred-mile trip from Salt Lake City to Houston, and the other half moving into a shared apartment adjacent to the campus, which he had found on a NASA employee classified-ads list the week before.

Less than ten minutes later, Thad found his way out of the tourist parking lot and onto an access road that led to the real JSC, which, as the kid in the space suit had correctly pointed out, was right next door. Thad knew for certain that he was on the right path when he caught sight of a mean-looking security gate barring the way ahead, attached to a rectangular kiosk with Plexiglas windows and floodlights hanging from each corner. Beyond the gate, Thad could make out a patchwork of asphalt lanes bisecting a pretty, rolling campus of green glades, low hedges, and boxlike buildings. He could also see objects in the distance that looked like oversized radar dishes and even a few buildings that reminded him of farm silos. Some of the buildings seemed modern, despite their 1950s-style exteriors, but a few of the complexes could easily have dated back to before the first moon landing.

The three oversized men in the security kiosk certainly fit the mid-century ethos: square-jawed, crew-cut, in gray-on-gray uniforms bearing the NASA emblem on the lapel and shoulder. As Thad pulled up next to the kiosk, one of the three leaned outside, holding out a beefy hand that looked more like a paw.

Thad rolled down his window and extended his driver's license, as well as the printed co-op acceptance form he'd received from Bob Musgrove in the mail, just a few days earlier. The guard took both, checking them against a list taped to the inside of the kiosk door.

"Welcome to NASA," the man said, handing Thad back his license. "Speed limit inside the campus is five miles per hour. Any faster, and we'll send a car after you. Oh, and since you don't have your ID yet, you'll have to park in the satellite lot. Once you're through the gate, go about a hundred yards and take a right at the rocket."

Thad looked at him. The guard grinned back.

"You're a co-op, right? That means you're supposed to be some sort of a genius. You'll figure it out."

With that, the man retreated to the crisp air of his kiosk, shutting the door behind him. He hit a switch on the security console and the gate swung upward. Thad pressed lightly on the gas, inching the Toyota up to five miles per hour.

The first stretch of road inside the JSC was fairly unremarkable, bordered on both sides by green grass, swampy-looking trenches, and some oversized hedges. But just as he made the first turn, Thad passed one of the high hedges and saw something through his windshield that nearly made him slam on the brakes. Stretched out along the entire right side of the road was a huge, cylindrical rocket. The thing was truly massive, over 360 feet from the rounded command module attached to the rocket's tip to the huge, jutting nozzles of the first of five fuel capsules that made up its cylindrical body. Blindingly white in the Texan sun, the thing could not be described as anything but beautiful.

Thad had read about the Saturn V rocket, but seeing it up close, just lying in a glade of grass at the opening of the JSC campus, was truly awe-inspiring. He remembered from his reading that this specific rocket had never been to space—it had been built for the two Apollo missions that were canceled at the end of the space race—but its sisters had carried every lunar orbiting and landing crew that had made the journey to the moon, and this beauty, though dormant and beginning to mold, had once been fully functional. In person, the scale of the thing brought home to Thad where he was: this wasn't some museum or amusement park; this was a place where real men and women trained to go into space, carried by machines such as the Saturn V. Christ, what it must have been like, being strapped to a beast like that, hitting speeds that made your skin compress right up against the bone—it was hard to even imagine. That this thing was real, not some model or mock-up, made the moment even more thrilling.

As Thad reached the end of the rocket, he saw a small group of people being led around the spacecraft thrusters by a young woman wearing a NASA uniform—some sort of tour, he assumed. The tourists all had badges on necklaces hanging around their necks. Thad also noticed that there were cameras along the road, a few trained toward the rocket and the tour group, but others aimed up and down where he was driving, covering what appeared to be every inch of the way in and out of the JSC. The cameras and the badges, like the rocket itself, again brought home the importance of where he was—that this was the first moment of an adventure that would surely change his life. He wondered how many people were watching his progress through the compound. Even from just inside the entrance, he could see that the place was huge. He knew there were over a hundred buildings on sixteen hundred acres—but those statistics didn't capture the open feel of the compound, how truly immense the place was. And this was just one facility of NASA; all around Clear Lake, the part of Houston where the JSC resided, there were multiple complexes serving the astronauts and scientists as they trained for future missions. And now Thad was part of all this, and would be for the next few years. He would be making a name for himself, impressing the people that mattered—and ultimately, always aiming toward his real goal of becoming an astronaut. Of maybe one day being the man who took that first step on Mars.

He was grinning hard as he finally pulled into the satellite parking lot and found a spot at the end of a row of cars sparkling in the growing heat. As he stepped out onto the pavement, he reached again for his sunglasses. His adrenaline was spiking as he squinted through the haze, making out the closest building where he'd be able to ask directions to the co-op orientation. It would be a bit of a walk in this heat, but he didn't care. He could already tell that he was going to love this place.

A few days ago, he was studying while helping Sonya fold laundry in some Salt Lake City Laundromat.

Now he was standing a hundred feet from a fucking rocket ship.

He almost laughed out loud as he took his first few steps toward the interior of the JSC campus.

5

The air-conditioning was jacked up so high Thad could almost taste the Freon on his tongue. Even so, rivulets of sweat trickled down the skin of his back as he stood near the end of a long line of young, pretty people snaking through the cavernous lobby of the industrial-era building. The walk over to Building 2 of the Johnson Space Center had been blistering, but as tired and hot as Thad was, he still had rockets on his mind as he waited with the other co-ops to enter the Teague Auditorium for the orientation lecture.

Locating Building 2 hadn't been all that difficult, even without the aid of a map; the swarm of smiling, highly energetic kids gathered around the front entrance would have been hard for Thad to miss. Now that he was among them, he could tell that his new colleagues represented a world he'd never been a part of before. Because of Sonya, and later, at the U of U, Thad had broken out of his shy shell, but he could still count on one hand the number of people he truly called friends. Friday nights, at best, meant a small dinner party and maybe a movie. He and Sonya were a self-contained unit rather than part of some definable social scene.

But here—packed into this refrigerated lobby full of young, exceedingly jovial kids—Thad was a blank slate. He could easily reinvent himself. Hell, the only person in the room who even knew his name was standing just inside the auditorium door, looking just as he had in the picture taped to Thad's living room wall—wearing a mischievous grin and a white shirt, joking amiably with the nearest co-ops. Bob Musgrove was welcoming the students one by one as they entered the auditorium. But other than to Musgrove, Thad was an unknown here—and in many ways, that thrilled him even more than the Saturn V rocket outside.

He was so caught up in thoughts of his own reinvention that it took him a moment to notice that the girl in front of him had turned half toward him, smiling. She was blond and tan, only a few inches shorter than Thad; her surfer-esque body looked fantastic beneath a white T-shirt and tight designer jeans. In fact, most of the co-ops were above average in looks; there was a preponderance of blondes with good figures.

The blonde in the T-shirt introduced herself as Sally Bishop, and after shaking Thad's hand, she pointed toward the wall behind him.

"That pretty much says it all, doesn't it?"

Thad wasn't sure how he'd missed the mural before, because it was utterly enormous. It took up an entire section of the lobby wall, painted in such bright colors that it competed with the near-nuclear glow of the high Texan sun streaming through the skylights above.

"I read about that mural in the orientation booklet," the girl said. "It's got some stupid name, *Opening the Next Frontier—The Next Giant Step*, but it's all right there. Instead of the orientation lecture, they should just have us look at the mural all morning."

Thad laughed. He'd also read about the sixteen-by-seventy-foot mural, painted by Robert McCall back in the seventies. It was supposed to tell the entire story of the JSC, from its birth in 1960 to the space shuttle program. The painting seemed a bit tacky, if not outright kitsch, but it did a pretty good job of graphically recognizing the space agency's accomplishments. From the first manned spaceflight of Alan Shepard in 1961, through the Gemini, Apollo, Skylab, and shuttle programs—Thad didn't think anyone who cared about space could stand in front of that mural and not get goose bumps. *Especially with the air-conditioning system blasting frozen air from every direction.* The thing that Thad liked most about the mural was where it ended; there was plenty of space along that vast wall for whatever came next.

"Maybe your picture is going to be hanging there one day," Thad replied. "I think you'd look pretty good in a space suit."

"It's going to be a while before any of us are wearing space suits. I'm just glad I made it here at all. Two days ago I was in Mexico with my boyfriend, and I forgot my passport in the hotel. I had to talk my way across the border. Good thing I had a bunch of mechanical engineering textbooks with me. The border guards had a weakness for a couple of NASA nerds."

"Your boyfriend is a co-op, too?"

"He's coming in later this afternoon, from Dallas. We're hoping to get assigned to a project together. Although I've asked around, and it seems like nearly everyone here is engineering."

Thad nodded. He was going to be in the minority, especially because he had listed geology as his main interest in the acceptance letter. He knew that after the orientation lecture, the co-ops were going to be assigned projects in areas as close as possible to their interests. It was just another thing that set him apart—hopefully in a good way. After all, how many engineers did it take to fly a spaceship?

As the line of co-ops slowly progressed into the auditorium, the girl continued her slightly flirtatious conversation. She told him about her wild trip to Mexico, about how awesome her freshman year at the University of Texas had been—UT being one of the five schools where most of the co-ops had come from—about how she'd fallen in love with the idea of working at NASA as a kid because her father, an ex–air force pilot, had forced her to go to space camp the summer after her sophomore year of high school.

For his own part, Thad began his reinvention by giving her only the abridged version of himself. He talked about how he and Sonya had recently become obsessed with paleontology, how he'd used his geology background to get them volunteer work at the university's museum. How they'd been invited on digs sponsored by the museum, and how fun it was to sift through the mud, chasing fossils, using science as a tool to re-create things they'd read about in books. Animated, he described how he'd found an actual *Tyrannosaurus rex* tooth on their most recent dig—only the fifth tooth found in that area of Utah.

Of course, he didn't mention the fact that while working as an inventory assistant at the museum, he'd also borrowed a few particularly cool fossils he'd been transporting to the storage closet—one rock jammed into his pocket becoming a few more fossils added a couple of days later—displaying them in his living room, often bringing them out at dinner parties to impress Sonya's friends. But he didn't think there was anything wrong with showing off such precious objects—the bigger crime, to him, was leaving those fossils in crates in a dark basement. Wasn't displaying such historical objects the whole point of a museum in the first place?

He had a feeling his fellow co-op would understand; she shared his adventurous bent. And listening to snippets of conversations going on all around him, he knew that the two of them were not alone. He was in a place full of young people with vivid spirits.

When he finally reached the auditorium, shaking Bob Musgrove's hand for the first time—getting a full pat on the back and a warm welcome—Thad was completely swept up in the emotion of the moment. He felt like he had found a home.

The feeling only grew through the introductory lecture, led by Musgrove and continued by a handful of JSC speakers. These speakers included a real live astronaut, in full uniform, porcupined with glorious, colorful NASA patches that marked him as someone who had flown in the shuttle—*actually been to space.* The astronaut detailed the history of the JSC—really, just giving life to the pixels in the mural hanging on the wall back in the lobby:

How it all started with a Russian dog named Laika: two months after Sputnik One stunned the world and put the fear of Soviet-controlled space in America's mind, the Russians managed to put a mutt, Laika, into orbit. No matter that the poor dog died from heat and stress on the way up—Eisenhower, terrified that Russia was going to win the space race, began plans for an astronaut program. In April 1962, construction of the JSC began in the Clear Lake area of Houston—a place chosen because of its smooth topography, and the fact that Rice University was willing to give the government a cheap lease price for the land.

The astronaut had the audience of co-ops enrapt from the very first word, although that could have been the result of his uniform and his natural cowboy swagger. He described how NASA moved from the Mercury program, which basically was about strapping men—equal parts brave and insane—to rockets aimed at low orbits, to the Gemini program, which was all about sustained life in space. Nine astronauts were chosen from a pool of almost eight thousand, given the name "The New Nine." They flew ten missions, the third of which, while launched from Cape Canaveral in Florida, was backed up by the newly finished Mission Control Center in Houston. It wasn't until Gemini Four that a full mission was controlled from the Houston center—made even more significant by the fact that it was also the first extravehicular space walk in human history.

From there, the astronaut continued into the Apollo period, briefly reliving the moon landing, the greatest accomplishment of the last hundred years. Thad tuned out as the astronaut cycled through the eleven Apollos that flew from 1968 to 1972. Like everyone else in the room, he'd seen the movie. He was more interested, for the moment, in scanning the crowd around him, the faces so filled with what could only be described as ecstasy. Even the description of Skylab, the least flashy phase in JSC history, didn't shake the elation from the audience. The story of how the Skylab space station eventually crash-landed in western Australia—causing a backwoods Australian city council to fine NASA four hundred dollars for littering—was just another parable in what could only be described as a story of biblical importance. To the co-ops, NASA was a religion. And a real live astronaut was nothing short of a deity.

The blue-suited man finished his lecture with the story of the birth of the shuttle program. On April 1, 1969, a group of engineers was told to report to Building 36. A NASA engineer entered the room carrying a balsa-wood model of an airplane, which he tossed toward the gathered men. They assumed it was a prank, but in reality it was an illustration of NASA's new direction. They were going to build a spaceship that flew like an airplane. By 1978, NASA was ready to elect its first group of shuttle astronauts, which they dubbed "The 35 New Guys."

The astronaut ended his speech to uproarious applause, and was followed onstage by Musgrove again, who told a few more jokes and then went through the actual details of the co-op program. Thad had already been through the rules booklet many times. He knew he'd signed up for at least three semester-long "tours"; after each, he'd have to return to Utah to continue his actual schooling. He'd be paid enough to cover room and board, maybe a little bit extra, but he doubted that any of them were there for the money—as evidenced by the fact that the co-ops were still mostly staring at the astronaut in his blue uniform at the edge of the stage rather than at the amiable man in the white shirt at the lectern.

"Keep your eyes open every day," Musgrove concluded from the stage. "Because every day in this place, you'll see something that's going to open your mind in ways you've never imagined. And maybe, if you work hard, if you're lucky, if we're all lucky—one day one of you will be standing here in a blue uniform telling us what it's like to walk on Mars."

Thad felt his face flush as he joined the other co-ops in applause. Musgrove finished by telling them to line up again in the lobby to receive their initial work assignments—but Thad was barely listening.

In his mind, he was already wearing that blue uniform, taking that first step on Mars.

6

Now we're talking.

Thad balled up a photocopied map of Building 31 into his fist, jamming it deep into his pocket, as he stepped across the threshold of the state-of-the-art Astromaterials Lab. Overpressurized, antiseptic-tinged air smacked him full in the face, and he grinned, taking in the three-hundred-square-foot lab with quick flicks of his eyes. He could tell immediately that he was in the right place. Glistening, stainless-steel counters, bucket-style, chrome-plated sinks, skyscrapers of test-tube racks, catacombs of Bunsen burners—and enough pipettes to build a church organ. The place was a scientist's wet dream, from the skating-rink-smooth cement floor to the achingly bright fluorescing panels that lined the ceiling. Even the overbearing hum of the level-four ventilation system seemed a symphonic throb in Thad's ears. This place put the geology labs back at the University of Utah to shame, and Thad could hardly believe he was going to be spending the next three months watching his reflection dance across all that chrome and steel.

Unlike Building 2, the Building 31 lab hadn't been easy to find. The place was a maze of windowless corridors and unlabeled doors. Because Thad was the only new co-op without an engineering background, he had been the only one assigned to life sciences. It was a cool distinction, because life sciences was interdisciplinary—which meant he was going to get access to a number of different labs in a variety of NASA complexes. He was going to be able to chase some really diverse interests over his three tours at the JSC, and if he played his cards right, there would be a lot of opportunity to work with and impress the higher-ups. The downside, however, was that it was another thing separating him from the herd. He would have to find his way on his own—just like he'd had to find the Astromaterials Lab, where he was supposed to spend his first few days, with little more than a poorly drawn map and a handful of directions given to him by Bob Musgrove.

But Musgrove and the map were erased from his thoughts the minute he stepped into the pristine, supercontrolled environment. He could imagine himself spending countless hours conducting experiments in this place, separated from the outside world by cinderblock walls built to withstand the strongest hurricane on record. In fact, he was so swept up in his own thoughts that he didn't notice the other person in the lab until he was almost right on top of him: a stringy young man around Thad's age, wearing a white lab coat over what looked to be blue scrubs, his hair covered by a matching blue surgical cap. The guy had his back to Thad and was leaning over one of the stainless-steel counters, a large rectangular object in his gloved hands.

Thad froze, staring at the object—because it was like nothing he'd ever seen before. It looked like a windowpane, but so incredibly thin—it didn't seem to have any depth to it at all. Not exactly transparent, but not opaque either—somewhere in between. Like fog, or a slice of cloud, somehow turned to glass.

"That's not something you see every day," Thad finally murmured, barely loud enough to be heard over the whirring ventilation system.

The young man at the counter didn't respond. Instead, he carefully placed the object down on a gel-like container, and then exhaled. After making sure the pane was secure, he turned to face Thad. Yanking the surgical cap off his head, he ran a hand through his unruly tufts of dirty-blond hair. His face was incredibly angular, his chin so sharp it looked like it was formed to cut stone. His jutting cheeks were bright red, and there were teardrops of sweat circumnavigating his pinpoint eyes.

"It's called 'aerogel,' and it's a bitch to work with. Lowest-density solid ever invented, strong enough to hold one thousand times its weight. And yet it shatters if you even look at it wrong."

"That sounds like a contradiction."

"Yep, pretty much sums it up. You make it by pulling all the water out of a silicon compound. It's an amazing insulator, but it weighs next to nothing. A piece the size of a human would weigh less than a pound and be able to support the weight of a car. If we ever really go to Mars, this stuff is going to be a big part of how we get there. And it's got a really awesome name. Liquid Smoke. How fucking cool is that?"

Thad grinned back at the kid.

"Very fucking cool. I'm Thad Roberts."

"I know, Dr. Musgrove texted me that you were on your way down here. I figured it would take you another ten minutes at least—you must be one of the smart ones. I'm Brian Helms. I'm going to be your lab mate."

Brian yanked off a glove and shook Thad's hand, then jerked his head to the left, indicating that Thad was to follow him toward another counter on the other side of the rectangular room.

"I'm a co-op, too, on my second tour. You really got lucky, man; astromaterials is the best gig here. We get to do just about everything. Especially now that everything in this place is all about Mars."

Helms reached the far counter and waved his one gloved hand at the objects strewn across the shiny surface. Thad saw various-sized rocks in containers ranging from petri dishes to strange, spherical globes that seemed to be filled with transparent liquid.

"This is what we do, mostly. Practice and experiment with preparation techniques, getting samples ready for transport to different locations around the JSC, as well as places outside of NASA."

"What sort of samples?"

"That's the really cool part. Up until now, it's been mostly lunar rocks. Or more accurately, lunar dust, because we're usually talking about a gram here, a gram there. But lately it's more about meteorites. Because some of those come from a lot farther than the moon—and that's what everybody's interested in now."

Thad looked at the various rocks splayed out across the steel counter.

"You mean some of these are moon rocks?"

"Of course not. Do you know how valuable moon rocks are?"

Thad shrugged.

"Actually, I don't."

"Very. Fucking. Valuable. And they have to be kept in really pristine conditions. You should see the Lunar Lab. We're talking Plexiglas cabinets filled with high-purity nitrogen. You go in wearing bodysuits, through these clean-air purification chambers—really sci-fi kind of shit."

Thad could only imagine what his new lab partner was talking about. He'd never worked with dangerous chemicals or biohazards before, so he only knew what he'd seen on TV, but he guessed it would be pretty cool to see the Lunar Lab in person.

"In this lab," Helms continued, "we practice on regular Earth rocks. You'll learn how to shave off little pieces, mimicking the ones from real lunar and meteorite samples that are often sent around to high schools as part of NASA's educational outreach program. I'll also show you how to put together a desiccator, which is a really cool device that keeps moisture out. For museums, we use these bigger glass spheres. They're usually filled with nitrogen to keep the rocks in good shape."

"So you pretty much run this lab?" Helms was just a co-op, but he seemed amazingly confident, like he'd been doing things on his own for a while.

Helms grinned, shaking his head.

"I'm just a wannabe like you. The division chief is Dr. Cal Agee. His assistant is David Draper. They're basically our mentors here in astromaterials. They'll come around now and again to make sure we're not setting the place on fire, or playing catch with the moon rocks. But just walking around the halls, you're going to meet a lot of scientists with as many letters *after* their names as you've got *in* yours. That's the best part of this place, hobnobbing with guys who play with space toys for a living."

"And we also get to work with astronauts?" Thad asked.

Helms gave him a sideways look.

"That a big deal to you?"

"Of course. I mean, scientists are cool, but astronauts are rock stars."

Helms laughed.

"I guess I'm a little jaded. Growing up around them kind of shakes some of the moon dust off."

"You grew up around here?"

"A few miles away. My mom is an engineer, did a lot of contract work with the agency."

Helms began removing his second rubber glove, struggling a bit to get it over his spindly fingers.

"A couple of days, and that wild look in your eyes will fade. You'll be having lunch, and suddenly realize that the guy sitting next to you once flew the space shuttle. And then you'll go back to your Fritos. Back to your textbooks and test tubes—and you'll realize this place is a job, as much as it's a dream."

Gloves off, Helms headed toward the door, gesturing for Thad to follow.

"We'll have plenty of time to talk about this later. You're going to the pool party, right?"

"Pool party?"

"Happens every couple of weeks. A few of the girls live in a complex with a common pool, and they throw pretty kick-ass parties. It's kind of a ritual. People get drunk, talk about things they maybe shouldn't. You'll find that this is a really social place—despite its stiff reputation."

Thad followed his new friend to the door. Even though he had been at college many years now, he hadn't attended too many parties. Partly because he was married, partly because of his upbringing; he'd never really gotten the hang of the whole party scene.

But then again, this was a place for reinvention. Sonya would understand. She'd want him to get the most out of this experience.

"A pool party sounds like fun."

"First we're going to go to the most important place on this campus."

"Where's that?"

Helms grinned, leading Thad out the door, making sure to shut it tightly behind them. Thad noticed there was a very high-tech-looking, computerized lock next to the door handle, consisting of a panel covered in raised, numbered keys. The minute the door shut, the computerized lock whirred, and a digital light began to blink.

"The cafeteria," Helms answered. "That's the real nerve center of NASA."

Thad was still peering back at the strange computerized lock. Helms noticed, and pointed with his thumb.

"You have to be careful, always make sure the cipher lock kicks in behind you. They take security really seriously here. You don't want to get yourself kicked out before you even start."

Thad raised his eyebrows.

"Then I'd never get to be an astronaut, right?"

Helms groaned. "Ha, you want to go to the moon someday?"

"No. I'm going to be the first man on Mars."

Thad wasn't even sure why he'd said it. He felt a little foolish, but Helms just shrugged.

"Maybe you will be," he said as he led Thad away from the door. "If I don't get there first."

7

The object was bright red and coming in fast—following a low, elliptical trajectory, spiraling as it went, spewing off droplets of clear, pearl-shaped liquid, like a comet tail painting its route through the electrically charged air. The speed of the thing was terrifying, and Thad had only a moment to get one hand up in front of his face—but it did him no good. The object went right past his fingers and collided square into his forehead. The impact knocked him back off his feet as a spray of ice-cold water exploded into his face.

"Skylab, baby! That's what happens when you've got a low orbit and too much gravity!"

Thad shook the water out of his eyes as he fought to regain his footing in the shallow end of the pool. Helms was standing about ten feet away, crouching behind a pair of deck chairs, a second water balloon in his cocked right hand. There were girls sprawled across both deck chairs, pretty and blond and wearing bikinis. The one on the right was the same girl Thad had met in line outside of the Teague Auditorium—Sally Bishop, of the boyfriend who still hadn't shown up, but was presumably on his way. The girl to Bishop's left was equally blond, but average height; there was something a little more natural and soft about the way her body filled out her flower-patterned bikini. A sunburst of freckles trickled out across the bare skin of her shoulders and arms, and when she laughed, the area around her blue eyes crinkled adorably.

"You refrigerated water balloons?" Thad coughed, shivering as the last remnants of the projectile trickled down the bare skin of his back. "That seems a little excessive."

"I never show up to a party empty-handed," Helms started to reply—but he was cut off by another balloon hurtling past from behind his left shoulder, arcing high above the pool, then exploding like a mortar a few feet from the oversized barbecue grill on the other side of the cobblestone patio. Thad looked up and saw that the second balloon had come from one of the balconies overlooking the pool. There were more co-ops in bathing suits up on the second floor, many clutching bright red plastic cups, presumably filled from the keg that dominated the grassy area on the other side of the barbecue. All told, Thad counted at least thirty people at the pool party—and more were still arriving. He didn't know what college parties were like, but this get-together was damn impressive.

His attention shifted back from the balcony as the freckled blonde slid off the deck chair and lowered herself into the pool just a few feet away. She brought her hands up behind her head, pinning back her flowing hair—and the motion did wonderful things to her bikini top. Thad felt himself blushing, and he shyly averted his eyes. Reinvention or not, he still had a long way to go before he was going to be entirely comfortable in a scene like this.

"I'm Lisa Daniels," the girl said. "I think I saw you this morning in line at Space Center Houston."

"Yeah," Thad said sheepishly, "that was me. I guess I'm not all that bright, because I probably would have wandered around there all day if some kid in a space suit hadn't pointed out that I was in the wrong place."

The girl laughed.

"Actually, I made the same mistake, yesterday. I came a day early to scope the place out. Thank God."

Thad loved the fact that he was in a program where a girl this hot was nerdy enough to show up a day early to what was essentially a glorified internship. He leaned back against the side of the pool as he watched

Helms and the other girl, Sally, slide into the water next to Daniels. Almost immediately, a handful of other co-ops joined them in the shallow end. Everyone was a little nervous, a little excited, and maybe a little too exuberant; at least Thad knew he was. For some reason, he really wanted to impress these people. If he wasn't going to be shy anymore, he wanted to shoot for the other extreme; he wanted to become the center of the co-op social scene, maybe just to prove something to himself, or maybe just to quiet that feeling that still plagued him, that he didn't really belong. He was twenty-three, he was married, and he had been kicked out of his house around the same time as these other kids were probably graduating from junior high.

"So this is a monthly thing?" Thad asked, shaking a piece of water balloon out of his longish hair. "Seems like it should be more of a weekly thing."

One of the co-ops who had taken over a deck chair laughed. He was a tall, athletic-looking guy wearing a starched polo shirt that had probably cost more than Thad's entire wardrobe.

"We're going to be working too hard to party every week," the polo shirt said. "I mean, at least those of us involved in rocket engineering."

Thad tried to pretend that it wasn't a subtle knock on his lack of engineering background.

"It's just that this place is so great," Thad continued. "It's a shame to waste it on a once-a-month kind of thing. We should incorporate it into our training."

Thad's mind was working fast as he noticed he had caught the attention of all the co-ops within earshot. It was a good feeling, being the center of interest, and his brain was quick enough to take advantage of the situation.

"What do you mean?" Daniels asked. Her eyes lingered on Thad a little too long.

"We can make it into a little game. An educational kind of game. Like a contest."

Helms was looking at him, and there was something cautioning in his eyes. But Thad ignored him, now on a roll:

"Each week, we'll identify the most awesome, incredible experience you can have here at NASA. We'll come up with the coolest thing that a person could get away with—and whoever does that thing by the next week, by the next pool party, he's the winner."

The athletic kid in the polo shirt leaned forward.

"The coolest thing—you mean, like get an autograph from an astronaut?"

"I was thinking we could be a little more creative," Thad said. "Like, the things we've read about in the co-op brochures. Maybe getting a ride on the KC-135—the Vomit Comet, that airplane that goes up and down so you get a few minutes of zero gravity."

"Or sneaking into the NBL!" Daniels nearly squealed. "You know, the Neutral Buoyancy Lab, the biggest indoor pool in the world, where the astronauts train—"

"I'm not so sure this is such a good idea," Helms interrupted, but one of the other kids pushed his way in, shouting out another idea.

"Someone could get into the Lunar Lab!"

"Or get a picture in a real astronaut helmet!" someone else suggested.

"Or sneak into Mission Control."

Thad was fighting back a grin as he looked at the excited faces all around him. He knew what he had to do to solidify his role. *He had to top them all.* But he wasn't going to just say something—he had to say it, and then actually *do it.*

"I'll tell you what I'm going to do," he stated, quiet enough so they all had to lean toward him to hear. "I'm going to get into Building Five, and get right up next to the Space Shuttle Simulator."

The place went silent. Thad could hear the water splashing through the filter all the way on the other side of the pool. All of the co-ops were staring at him. Then the kid in the polo shirt laughed out loud.

"No fucking way. You'll never pull that off. Only astronauts get near the Space Shuttle Simulator."

The kid in the polo shirt was probably right. The simulator was more than just a mock-up of the space shuttle, like the one Thad had seen at Space Center Houston. It was a working, hyperrealistic flight simulator that you got inside and controlled like the real shuttle. Building 5 was one of the most secure buildings on the NASA campus. But Thad had already started down this path, and he certainly wasn't going to back down now.

"I guess we'll see." He shrugged. "But I hope you'll all be back here next week so we can talk about it."

After that, the party started to break up, and the co-ops drifted away by ones and twos. The freckled girl, Daniels, lingered as long as she could—but Thad did his best not to give her any special attention, because he really didn't want to lead her on. Sonya was a long way away, but Thad figured he had survived the Mormon Church for nineteen years; he could get through three months of pool parties and freckled girls in skimpy bikini tops.

He started to climb out of the pool when he noticed that Helms had moved close to him, still carrying that cautioning look in his eyes.

"Don't do anything stupid, man. You don't need to impress anyone."

Thad glanced at him.

"I'm not trying to impress anyone. I just think it will be fun. Lighten up. I'm not going to do anything that's against the law."

Helms looked at him a moment more, then clapped him on the shoulder.

"You keep this up, you're going to end up shaking this place up. I guess it could use it."

They both glanced over at the girls—Bishop and Daniels, who were moving past the deck chairs. Daniels adjusted her top to cover a little more freckled skin as she went.

"Still," Helms continued. "Like I said before, you don't want to get yourself kicked out of here. There's just too much goddamn opportunity."

Thad wasn't sure whether his new friend was talking about the girls or NASA—but he couldn't help but agree on both counts.

8

Smile for the cameras.

Thad kept his head low as he strolled by the entrance to Building 5 for what had to be the sixth time; he'd been casing the place for the past twenty minutes, but he still hadn't come up with anything resembling a plan. Christ, what an awful criminal he'd make. If anybody was monitoring the dozen or so cameras that were perched along the tree-lined path circumnavigating the modern-looking complex on the south corner of the JSC, they'd probably think that one of the co-ops had gone batty: a kid with bright green eyes in a NASA shirt and khaki pants, aimlessly circling one of the most secure astronaut training facilities on the campus.

Thad couldn't imagine what he'd been thinking when he'd made the boast at the pool party. He'd been at NASA less than two days, and here he was. Twenty minutes ago, he'd finished an amazing day spent learning how to slice meteorites into tiny segments; now he was wandering around a secure building, contemplating a stunt that could get him kicked right out of the program—and possibly even arrested.

As he passed a set of bushes that marked the far corner of the building, he fingered the yellow security badge that was hanging around his neck. Yellow meant level two—a step beyond what most co-ops wore, because of his position in the life sciences department. The badge meant he could get into most of the buildings on campus—but there were a few important exceptions. Building 1, where the NASA brass had their offices. Building 31N, where all the valuable lunar materials were kept. And Building 5. Yet here he was, wandering back and forth in front of the smoked-glass entrance, trying to figure out a way to get inside.

Helms had spent most of the day trying to convince Thad that it was an extremely bad idea. Helms had grown up in this environment; he knew how seriously people at the JSC respected rules and regulations. But to Thad, Helms was looking at it all wrong. Science was about going beyond the rules, taking chances. Wasn't the whole point of the co-op program to expand his mind? And besides, Thad only had to close his eyes and he could still picture himself in that swimming pool, all those other kids hanging on his every word. He'd never really felt like that before. And he wanted to feel that way again.

He stopped at the next set of hedges and doubled back toward the glass entrance to the building. There was nothing wrong with at least giving it a try. All he wanted to do was get inside the building, take a quick look at the simulator, then get the hell out. If he got caught, well, he could just play dumb. As long as no one looked at his résumé, he could probably get away with it.

He reached the steps that led up to the black glass entrance—but this time he didn't pause. He did his best to control his breathing as he entered the building.

He quickly found himself in a spartan foyer, facing a huge metallic door. The surface of the door was completely smooth; no knobs, buttons, or levers that he could see. Above the door was a single security camera pointing down at him. To the left of the doors was a punch pad and a tiny TV screen.

Thad felt a wash of panic spread through him, and he almost turned and ran back through the glass entrance. But he realized that the camera could already see him; before he could do anything, a woman's voice echoed out of the TV screen.

"Can I help you?"

Thad had to improvise.

"Yes. I'm here for the Shuttle Simulator."

There wasn't even a pause. A buzz echoed from behind the door, followed by a mechanical click. The door sprang open a few inches, and Thad quickly pushed through. He immediately found himself in a busy hallway. There were people everywhere, some in white lab coats, some in NASA shirts like his own. But Thad's eyes quickly focused on the astronauts—at least three that he could see—in their standard blue uniforms, all with shuttle patches on their shoulders.

Christ. Thad could feel his heart pounding in his chest. Again, he fought the urge to turn and run back the way he had come. But he'd gone this far—and the thing was, nobody in the hallway seemed to notice him. There were people everywhere, but none of them were paying him any attention.

He leaned back against the hallway wall, contemplating his next move. He saw that a few of the people in the hallway were holding clipboards. The guys with the clipboards seemed the least aware—so he figured that one of them was his best bet.

He waited until one of the clipboards moved past him, and then fell right in step behind the man. When the guy finally looked up, Thad smiled at him, forcing the nervousness out of his voice.

"Can you point me toward the Shuttle Simulator?"

The guy didn't hesitate at all.

"Sure, I'm headed that way."

Thad almost had to skip to keep up with the man's pace as he navigated his way down the long hallway. They took a hard, ninety-degree turn—and suddenly they were in front of another metal door. There was a card reader next to the door, and without pause the man with the clipboard took out an ID and swiped it through. He pushed the door open, leaned his head in, and hollered at someone on the other side.

"This guy's here for a simulation run, can you take care of him?"

Thad nearly choked. He wanted to say something, but his voice was completely gone. The man with the clipboard held the door open for him, and Thad had no choice but to step inside.

Oh, shit. Thad didn't even hear the door click shut behind him. He was standing at the edge of what looked like an enormous airplane hangar. There were computers everywhere, workstations separated by engineering panels and whiteboards, all of it interconnected by spaghetti snarls of thick black electrical wire. And there, in the center, rising high into the cavernous space, stood the Space Shuttle Simulator. It was nothing short of spectacular.

"Your first time? Wish it was always this easy to spot a virgin."

The voice came from Thad's left, and he glanced over toward a pair of technicians in matching light blue smocks, hovering over something that looked like an oversized circuit board. The one who had spoken was grinning at him, so he grinned back—but he couldn't keep his focus very long. Like a set of house keys in an MRI machine, his gaze was yanked back toward the technical wonder that took up most of the hangar in front of him.

"It looks a lot bigger in person," he mumbled.

The simulator was made up of two separate parts. The smaller of the two, the motion-based crew station (MBCS), as it was called, was attached to a huge scissor crane—a jointed, steel monstrosity, loaded with springs and curled-up pneumatic hoses, that assumedly provided incredible levels of hydraulic lift. The MBCS looked like the nose cone of the shuttle, gripped by a massive robotic arm. Although Thad couldn't see inside the thing from where he was standing, he knew that the MBCS was configured just like the real cockpit of the actual space shuttle, with room for the shuttle commander and the shuttle pilot. The arm gave it six degrees of motion—which meant the thing could simulate every phase of spaceflight, from launch to landing. It could tilt up to ninety degrees in every direction and could simulate acceleration, even moments of weightlessness.

The second part of the simulator was the fixed crew station. A rectangular box, it was a veritable porcupine of wires, antennas, and even miniature radar dishes. The MBCS had room for a commander, a pilot, a mission specialist, and a number of other crew members. It wouldn't simulate motion, but it was also raised up on an elevated platform, and it was supposed to perfectly simulate the interior environment of the shuttle itself. For long-duration mission simulations, crew members could spend days or even weeks in the MBCS. Food and water would be raised up to them so that they could live exactly like they would in an orbital environment.

"That's what one hundred million of your tax dollars will get you," the technician responded as he finally stepped away from the circuit board and approached Thad. "I assume you're here for the monthly systems check?"

Thad looked at the guy again. The technician was in his mid-thirties, with a receding hairline and a few extra pounds hanging down above his belt. Probably a contractor, obviously not someone Thad would consider an authority figure. No doubt the tech had confused him with someone who was supposed to be there. Or maybe he just didn't care. He saw the NASA shirt, and that was enough.

For a brief moment, Thad considered ending his charade. Something felt wrong about the deception, even though he hadn't actively done anything to convince anyone he was supposed to be where he was. At the same time, Thad couldn't ignore the spikes of pure adrenaline that were ricocheting through his system. It was like the first time he'd flown a single-engine plane by himself, but even more intense. He felt really alive, and the fear of getting caught no longer crossed his mind.

"That's correct," he heard himself respond. "I'm supposed to observe the test run."

"The rest of your crew is already inside," the tech responded, starting forward toward the simulator. "If we hurry we can make it before it begins."

Thad's eyes widened. He had assumed he'd be observing the test run from where he was standing. *Well, in for a penny, in for a one-hundred-million-dollar simulator.* There was no turning back now. He quickly followed the man toward the massive machine.

A second later, he was a few feet away from the giant hydraulic crane. The MBCS's nose cone was right in front of him, and the tech headed for an open hatch affixed to one side. The tech pointed through the oval opening.

"You guys have the coolest fucking toys."

Thad wasn't sure he was even breathing anymore as he stepped past the tech, bending his head so he didn't hit himself on the simulator's ceiling. Before he could blink, he was inside the cockpit of the space shuttle. At least, a mock-up so realistic no astronaut in the world would be able to tell the difference.

In some ways, it was like the interior of an airplane. Except a million times *more.* There were triangular viewing windows ahead, windows on either side—and every other surface of the thing was covered in switches, diodes, buzzers, and levers. There was already a man strapped into the pilot's seat to Thad's right. He couldn't tell if the man was an astronaut or a technician, because he was wearing what looked to be gray-on-gray overalls. But there was no doubt he knew what he was doing. His hands were flicking around the switches, beginning what had to be the launch sequence. Without looking up, he gestured toward the other chair—the commander's seat.

Thad felt another moment of extreme panic, which he quickly swallowed down. As he told the tech, he was just there to observe. That was the charade he had invented, and that was the charade he was going to stick with. Just a lowly co-op who had been sent by his mentor to witness the monthly check of the Space Shuttle Simulator.

It took a moment to figure out how to strap himself into the commander's seat. There were seat belts coming from every angle, and a holster that went around his chest. When he was done, the pilot said something into a communicator attached above their heads, and Thad heard the whoosh of the hatch sealing shut behind him.

"Let's finish the checklist," the pilot grunted, and Thad quickly looked where the man was pointing.

There was a printed checklist attached between their seats. Because Thad had his pilot's license, he was at least barely able to follow what was going on. He didn't know where anything was located, but he was able to mimic the pilot's lead, flicking a switch here and there, reading an alternator or a temperature control.

"Fire it up," the pilot said.

And the next thing Thad knew, the entire cockpit began to shake. At first, it was a low tremble, but then the thing was really jerking up and down, like a paper airplane riding across the top of a thunderstorm. And suddenly the whole cockpit tilted all the way on its back, nose pointing up. Thad stifled a gasp. To his surprise, the window ahead of him no longer looked out on a converted airplane hangar. Thad was looking at the sky. They weren't windows; they were high-definition monitors, playing feedback from a real shuttle launch.

A second later, Thad was slammed hard into his seat. The view through the windows became one of pure motion, streaks of light like laser beams flashing before his eyes. The noise of the engines was like thunder reverberating around him in truly deafening peals.

Thad realized he was shouting, in pure unadulterated joy. Maybe the pilot noticed, maybe he couldn't hear over the din of the mock thrusters—Thad didn't really care. In his mind, he wasn't in a simulator tucked into a secure building on the JSC campus.

In his mind, he was sitting in the cockpit of a rocket ship, hurtling toward Mars.

. . .

It wasn't quite the overwhelming energy rush of a simulated shuttle thruster pushing him back into a leather commander's seat—but it was pretty damn close. Sitting cross-legged at the edge of the same swimming pool from the week before, half the young population of Clear Lake spread out across the patio in front of him as he told the story—so many eyes and ears and minds focused entirely on him—maybe embellishing a little bit here and there, but keeping to the narrative as much as possible . . . well, it was a truly pivotal moment in Thad's life. He could see his own charisma reflected in the eyes of the pretty girls closest to him, and even in the unabashedly awed expressions on the faces of the men.

"So all in all . . ." Thad finally wrapped up the story. "I think it was a pretty good week."

There was a moment of frozen silence, just like there had been when he'd first proposed the idea of the contest a week ago. And then everyone was applauding at once, congratulating him, wave after wave of handshakes and pats on the back, and even a few kisses on the cheek. Helms gave him a grudging thumbs-up, shaking his angled head in admiration.

Thad had secured his place at the top of the social food chain. It was a spot he'd never occupied before—and he liked it.

When the crowd moved away, Helms sidled next to him, dipping his finlike feet into the cool water of the shallow end of the pool.

"Your contest was quite a success. I think it might end up a weekly thing. But I doubt anyone's going to top flying the space shuttle."

"It was just a simulation." Thad laughed. "I'll probably wait till my third tour to try and sneak into the real thing."

Helms laughed back—then paused and looked at Thad.

"You're kidding, right?"

Thad slid forward into the pool, submerging himself all the way down to his bright green eyes.

9

It was a moment every true scientist knew well—although it wasn't something quantifiable, it wasn't something you could predict or reverse-engineer or data-map or even really describe—but it was a moment that anyone who had spent time sequestered in a lab or behind a computer screen or at a blackboard, chalk billowing down in angry stormlike clouds, could identify, if not define.

Thad has his own word for it: *serenity*. The moment when the *act* of science organically shifted into the *art* of science; when even the most mundane, choreographed procedures achieved such a rhythm that they became invisible chords of a single violin lost in the complexity of a perfect symphony. Minutes shifting into a state of timelessness, where the world seemed frozen but Thad was somehow moving forward: content, fulfilled, free.

The project itself was far from spectacular. Slicing away at a piece of volcanic rock using a tiny diamond-tipped saw while keeping track of every microscopic wisp of volcanic dust—accurately documenting the final weight of the sample that was left behind. The work was painstaking, but the volcanic rock was just a stand-in, like the mocked-up cockpit of the space shuttle. It was supposed to represent something infinitely more valuable. A chunk of the moon, hand-delivered more than thirty years ago by men whose names were enshrined in history books. For

Thad, it didn't matter that the procedure was little more than a dress rehearsal. The process itself had overtaken him, and in that moment he was truly lost in the art of the science. The whir of the diamond saw, the pungent scent of the heated volcanic sample, the swirl of the dust as it billowed upward into a mercury-based measuring machine. He was in that serene place where nothing else existed. And he would have been content to stay there forever.

"Wow. You did all this by yourself?"

It took Thad a moment to process the words, to let the familiar voice yank him back into the lab. He switched off the saw and glanced back over his shoulder. Helms was standing by the counter where Thad had laid out all of his practice samples; everything from minute educational slices in individually wrapped Teflon bags to carefully constructed desiccators holding mock meteorites, ready to be sent to labs all over NASA.

"I wasn't sure when you were going to be finished running errands for Dr. Draper. So I figured I'd get started on my own. I guess I lost track of the time."

"I'll say. I assume Dr. Agee showed you how to do all this?"

Agee, Thad's official mentor, had indeed stopped by earlier that morning to introduce himself, but he only stayed for a few minutes. Thad had been on his own most of the day. That didn't bother him; actually, he found it quite liberating. His adventure at the mock space shuttle had taught him that NASA was a place an independent mind like his could take great advantage of. And Thad had become very independent. Ever since he'd been kicked out of the hermetic world where he'd grown up—the Mormon Church the way his father interpreted it, the heavy-handed way of the Mission Training Center—he'd become hungry to make his own future, to build his own name. The cool thing about the co-op program seemed to be that he'd be able to find his own way, to a large degree.

"He gave me some pointers. But I learned a lot of it from reading your notebooks, and using the checklist I found on the laptop."

Helms glanced back at the computer station in the far corner of the lab. There was one laptop, a couple of desktops, and some wiring that led into NASA's secure mainframe. It was a pretty high-tech station, and it was also highly secure. Helms had duly informed Thad that NASA security could monitor any use of the computer system, including personal e-mails. Thad figured that was for the best. Even a cursory search of the mainframe using the laptop had told him that there was a lot of pretty cool information available to an employee with his level of security. He could only imagine what higher security clearance would get you.

"You missed lunch," Helms said, moving next to Thad to help him begin to disassemble the saw. "But if we're quick, we can grab something on the way over to the lecture."

Thad raised his eyebrows. He really had lost track of the time. He'd been at NASA over a week now, but he'd only made it to the Stardust Café twice. He didn't really care—food had never been a real priority for him. Back at home, Sonya had often had to remind him to eat. As a struggling model, she had found the ease with which he skipped meals quite annoying. But certainly at NASA, no matter who you might run into in the cafeteria, Thad found meals the least interesting part of his day.

The upcoming lecture was a perfect example. Although Thad had yet to meet Dr. Everett Gibson, he certainly knew the man by reputation. Gibson had been a standout scientist in the life sciences division for well over thirty years. One of the brightest stars in the astromaterials research office, he was the epitome of the old-school NASA scientist. After attaining a master's in physical chemistry and a Ph.D. in geochemistry, he had gone to work for the JSC—or, as it was called at the time, the Manned Spacecraft Center—in July of 1969, just as the Apollo 11 capsule first returned from space.

Thad was already in awe of the man. Gibson wasn't an astronaut, but he was the closest thing a laboratory scientist could ever dream of becoming. It was fitting that he would be giving his lecture in the life sciences building, where Thad worked. Gibson had spent almost all of his thirty-plus years at NASA stationed in that building because—as Thad had learned only the day before, during a bull session with Helms and a couple of other co-ops—Building 31 had once housed the lunar receiving laboratory.

When the first few Apollo missions had come back from the moon, NASA had set up a really tight quarantine; nobody had any idea what the lunar samples they had brought back might contain. There was a very real fear of alien pathogens spreading some strange, unearthly disease throughout the space center—and perhaps from there, the entire world. So a high-tech quarantine had been created not just for the astronauts themselves, who spent weeks in sealed chambers going through multiple levels of purification, blood tests, and even psychological evaluation, but also for the lunar samples—the moon rocks, as they quickly became known to the public.

The protocol for the transport and storage of moon rocks was unbelievably strict, involving vacuum-sealed rock boxes, nitrogen chambers, bodysuits with self-contained oxygen.

Gibson was one of the first scientists charged with preparing and studying the moon rocks brought back by the Apollo program, missions twelve through seventeen. He had conducted the original moon rock studies, searching for signs of life, unknown materials, pathogens—everything that made the lunar samples unique. Eventually, the quarantine on astronauts and materials was lessened, and by Apollo 15 dropped. The rocks, though deemed incredibly valuable, were no longer considered a danger. But they were still irreplaceable; after the Apollo program ended, it was immediately made illegal for American citizens to even own a real lunar sample.

All together, the Apollo astronauts had collected 842 pounds of the stuff, divided into 2,200 individual samples, which were then subdivided into 110,000 studiable parts—and it had been determined that the moon rocks needed a building of their own. A self-contained facility, Building 31N, had been constructed right next door; Thad had yet to visit the Lunar Lab, but he had heard plenty of stories about the place. It was considered the most secure building NASA had ever built. Atmosphere-controlled, built without any connection to the outside world—no wires, pipes, ducts—it was supposedly strong enough to survive a thousand years underwater without damage to the inner contents. Hell, it would probably outlast the entire city of Houston.

Thad hoped he'd get the chance to visit the Lunar Lab. Since he was involved in the study of space materials, he knew that it was not far-fetched. But until he got to handle the samples himself, the closest he would get to moon rocks would be hanging out with people like Dr. Gibson.

"Forget lunch," Thad said, hastily cleaning up his workstation. "I'd rather starve and get a good seat up front than be bloated in the back row."

Helms smiled, though, as a second year, he'd heard Gibson's lectures before. But nobody in Building 31 missed an opportunity to hear about the Apollo missions from a man who had been so involved himself. It was as close to walking on the moon as a guy who worked with test tubes was ever going to get.

. . .

Gibson began his speech on the moon, but the body of his talk took them millions of miles beyond; sitting in the very front row of the Greek-style amphitheater, leaning all the way back so that he had a better view of the stocky, square shouldered, sixty-something-year-old man behind the lectern, Thad realized he should not have been surprised. Like everything else at NASA, Gibson was caught up in the incredible reorientation of the American space program. But it was still amazing to see this gray-haired, slightly balding, bespectacled scientist—a genius who had taken part in the greatest adventure in modern human history—so enthusiastically involved in something new, something that would take at least a quarter of a century to become real.

At the beginning of the speech, Gibson talked about the first samples he'd ever seen when he started at NASA—the Apollo 11 samples, which were collected by Neil Armstrong, the first man on the moon. Gibson went on to describe how different the samples were from each successive moon mission, how each of the six landing spots had been chosen to study different areas of the moon's topography. The results were startlingly different types of rock, from the very fine-grained materials brought back from the valley of Taurus Littrow, a deep mountain valley in the northeastern part of the moon—a material that was made up of little tiny beads commonly known as "orange glass," to the gray, almost black Apollo 17 samples—the very last samples ever collected by human beings, from the dark portion of the moon. In passing, Gibson acknowledged how utterly valuable the samples were—not just that they were irreplaceable, but that requests came in from all over the world, every day, from scientists, museums, and colleges wanting to display or study these national treasures. And every year, NASA chose a few hundred lucky souls who would have a chance to see a moon rock for real.

Thad smiled as he listened to this part of Gibson's speech. In his lab, just a few floors away, he had been practicing the techniques that would be used to prepare those moon rocks. He was part of the NASA machinery, part of the fraternity of scientists who made such science possible. When Gibson added, almost as an aside, that these samples were also infinitely valuable in nonscientific terms, Thad barely registered the thought. That someone had once tried to sell a single gram of illegal moon rock for $5 million—that really didn't mean anything to Thad, at the time. The value of those lunar samples went well beyond money. They represented the greatest human endeavor in history.

Once Gibson shifted his talk away from lunar rocks, Thad did not believe the man could somehow refocus the audience's attention, but then Gibson shocked them all by reaching behind the podium and lifting up a small glass vial. From the front row, Thad could actually make out what was inside—a glassy-looking piece of rock, almost volcanic in nature, but certainly something that he had never seen before. Gibson smiled at the crowd as he said the object's name:

"ALH 84001. Recovered from the ice in Antarctica in 1984, this little thing has been the focus of my life for nearly a decade now. In 1996, I published the scientific results of my studies in *Science* magazine. I'm sure some of you have read it. This meteorite, which is over four billion years old—we believe it came from the planet Mars. And this meteorite contains within it evidence of past biological activity. In other words—this meteorite suggests, unequivocally, the possibility of life on Mars."

Thad reacted with the rest of the audience, awed and amazed. He glanced around himself, saw the raptured faces of the co-ops around him; it was one thing to impress a swimming pool full of college-age kids with a story about a trip on the Space Shuttle Simulator, Gibson had shown an entire amphitheater evidence in support of life on another planet. The man had held in his hands moon rocks from every landing in human history—and here he was, holding a piece of Mars itself, dredged up from the deep ice of Antarctica.

Thad may have been in the process of reinventing himself as a social leader of the JSC co-ops, but Everett Gibson was a fucking rock star.

. . .

After the lecture had ended, and the audience had filed away toward the various labs, cafés, and workstations that dotted the NASA campus, Thad lingered behind. He waited until Gibson had packed away his notes—and the Mars sample—into his leather-bound, NASA briefcase, before approaching the edge of the stage. Helms was a couple of rows back, chatting up a pretty coed from the University of Texas. Even so, Thad could see, out of the corner of his eye, that Helms was partially watching him. Helms, it seemed, was always watching out for him, maybe worried that Thad had the capacity to push things too far, take too many chances. Thad was amused by the thought. NASA was a dream come true; he had no intention of ever doing anything to ruin that dream.

All he wanted to do was introduce himself to the man who had just opened his eyes. Eventually, Gibson noticed Thad at the edge of the stage. Gibson strolled over, his gait casual, if a little stiff. He leaned down so Thad could shake his hand.

"I'm Thad Roberts. I work in the building. I'm on my first tour."

"I recognize the name," Gibson said, smiling amiably, "and I look forward to getting to know you better during your time here at NASA. You enjoying yourself so far?"

"I feel more alive than I've ever felt," Thad started, realizing he was talking way too fast, maybe even bordering on a speed that could be described as manic. But he couldn't help himself. "I can't believe I'm shaking hands with the man who discovered life on Mars."

"Science is a group effort here at NASA. A lot of brilliant people put in a lot of time to make discoveries like this, as I'm sure you'll learn. Nothing happens here overnight. And it's more important to be part of a brilliant constellation than try and go it alone."

Before Thad could respond, a group of older co-ops sidled in front of him, capturing Gibson's attention. Thad found himself bumped and jostled away from the edge of the stage—until he was nearly back at the front row of seats, watching the brilliant scientist holding court. He could still feel Brian Helms's gaze on his back, but he didn't yet turn around.

Deep down, he understood what Gibson had told him, that being part of a brilliant constellation was what a place like NASA was all about.

What Thad didn't realize yet—but would soon learn—was that being just one bright star in a constellation simply wasn't in his nature. Men like Everett Gibson—and Brian Helms—could be content being shiny parts of this historic solar system. But Thad would always need something more.

He wanted to be the brightest star—the one everyone saw when he or she looked at the sky. And the scary thing was, it didn't really matter if that star was bright because it was the biggest—or because it was just about to go supernova.

My Dear Rebecca,

The seasons have swayed back and forth, but time has flowed beyond me—left me alone as I gaze into the horizon where your memory still shapes the entire world. The wind blows in the other direction constantly interrupting me, the rest of the world has moved on, all things are destined to decay. But the wind has never known its center, it dances out the curse of ever-grasping, always passing everything by. All it knows is the sad song of moving on.

I once knew a beautiful young woman that didn't believe in forever. She became my forever.

10

"Ten. Nine. Eight. Seven—"

"Look at his eyes! He's not going to make it!"

"Six. Five. Four!"

"Come on, man, almost there!"

"Three—"

Thad let out a sudden howl. His right arm shot forward, almost involuntarily—and he desperately grabbed for the oversized yogurt shake in the center of the round wooden table. He was halfway out of his chair, his other hand gripping the edge of the table so tightly his fingers were almost the same color as the shake. His knees were trembling, and there were beads of sweat running down his bright red cheeks. He nearly knocked over the massive platter of tandoori chicken as he fought to pilot the shake toward his blistering lips.

"And I really thought he was going to make it," someone groaned.

Thad collapsed back in his chair to a chorus of applause, laughter, and even a few catcalls. There were only six people at his table in the back corner of the tastefully, if somewhat inauthentically, decorated Indian restaurant, but a few of the other nearby tables had joined in on the fun. No doubt everyone in the place—from the group of women seated at a ridiculously ornate, canoe-shaped bar in the far corner of the restaurant to the scattered parties of middle-aged men at the imitation hookah stations throughout the interior of the brightly curtained Indian dive—was a NASA employee of one sort or another. One week into his second tour, Thad guessed that some of them recognized him, some of them didn't. But certainly they knew the other faces at his table, because Thad was in the process of being tortured by some of the greatest minds of the modern day.

"Don't worry about it, kid," the man directly across from him said with sympathy as Thad took large gulps of the strange, unpronounceably named yogurt. "I had Sanjay mix something a little special into the tandoori. I don't even think Saumya could have gone ten seconds—and he was born with a hookah in his mouth."

A fifty-something Indian to Thad's right slapped him on the shoulder.

"It was a noble effort. And you'll have a chance to redeem yourself in the next course. If you thought the tandoori was hot, wait until you taste the bhindi masala. It is simply outrageous."

Thad laughed, despite the tears pooling in the corners of his eyes from the spices. The Kashmir Express was really a terrible restaurant, but for some reason it had become the favorite weekly meeting place for a large faction of the JSC elite. Certainly, proximity played a factor, but there were many—and far better—restaurants within five miles of the sprawling campus. Something about the kitschy, overspiced mock-up of a Bombay hot spot, located in a desolate corner of the South Houston strip mall, somehow connected with the lab rats who shared Thad's predilection for adventure.

The ritual of the Monday lunch had certainly not been Thad's idea; he was just thrilled to have worked his way to an invitation from such a prestigious crew. All of the men at the circular table worked in Building 31, but Thad was the only co-op in the group. Most of his interactions with these prestigious scientists had taken place in the hallways, elevators, and stairwells of the life sciences building. But even so, Thad had impressed them and—beginning the last month of his first tour, and somehow carrying over through the three months he had spent back at the University of Utah—he had become part of the weekly lunch crew.

If Thad had once felt like he didn't really fit in with the other co-ops—he was little more than a shadow on a wall with this crowd. The six scientists spanned a multitude of disciplines—geology, physics, astronomy, engineering—but the thing that bonded them was that they were all world-famous. If Thad had walked into any bookstore in the country and looked up popular books on Mars, he would've found their names. Although Everett Gibson wasn't at this particular lunch, he had joined the group at the Indian restaurant a number of times before. These were Gibson's colleagues, his contemporaries, and Thad—twenty years younger than the youngest among them—felt utterly blessed to be invited, even while they were torturing him with potentially deadly levels of Indian combustibles.

The cool yogurt finally doused the invisible flames terrorizing the membranes of his mouth, and Thad refocused on the spirited conversation that enlivened the table. Usually, the Monday lunch session was an opportunity for the scientists to try to impress each other, but today one of the more senior geologists from Building 31 was trumping them all; he had just returned from an Antarctic rock-collecting mission, the very sort of mission that had led to Everett Gibson's work on the Allan Hills meteorite.

"You'd be surprised at how upscale base camp is becoming," the geologist was saying, gesticulating with a piece of chicken. "Hell, I've stayed in worse places in New York. It might be thirty below, but at least there's room to stretch out."

As the man spoke, Thad could see himself swaddled in the latest state-of-the-art snowsuit, roving around the ice floes on a souped-up snowmobile, scouring the glacial plains searching for meteorites. He knew that the Antarctic glacier was basically a big ice sheet that worked like a conveyor belt. Snow fell and collected in the mountain areas, moved down into the ravines, pushing more and more rocks into natural collection areas. The NASA geologists made annual journeys to search these natural collection areas, where they were more likely to find meteorites that had fallen thousands, and even millions, of years ago.

The geologist at the table hadn't found anything quite as significant as Gibson's evidence in support of life on Mars, but he had come home with a pair of small meteorites that would lead to a few pretty good papers in the scientific journals. The other men at the table were clearly envious, but in a good-natured way. Like Gibson had said, NASA was about the constellation, not the stars.

After the man had finished, the conversation shifted in Thad's direction. Like the Indian locale itself, Thad's co-op reports had become an important part of the ritual. Just as he lived vicariously through the stories told by the elite scientists, the older men lived vicariously through Thad's experiences with the youngest members of the NASA family.

Beginning shortly after the pool party, Thad had begun arranging excursions for his co-op class. Growing up in backwoods Utah, he had always felt most comfortable outdoors—and he learned very quickly that the other NASA co-ops had very little experience with nature. So Thad had morphed into a de facto social chair, arranging adventures that ran the gamut of his expertise—from rock climbing to bungee jumping, and everything in between.

"This weekend wasn't anything special," Thad began, underplaying it just right. "Only a little skydiving down around the Galveston shore . . ."

He embellished as he went, not that it was really necessary. He'd found that as impressive as the scientists were, it was easy to thrill them with stories—as long as he included plenty of scenes with voluptuous first- and second-year co-ops, and a little bit of life-threatening adventure. Rock climbing, scuba diving, hang gliding, whitewater rafting—all things Thad had been enjoying for years with Sonya, but were totally new experiences for the JSC employees.

Everett Gibson seemed especially charmed by these tales of adventure. Thad remembered how the man had positively lit up when he had described what it was like to line up next to two terrified teenagers—who'd never jumped off the lowest branch of a tree, let alone off the top of a thirty-foot cliff—for a dive they'd remember for the rest of their lives.

In fact, most of the co-ops had never even been camping before, let alone cliff diving. Thad had given the interested ones a short lesson in survival skills, everything from starting fires to climbing trees. He had thought it was ridiculous that these brilliant kids had never done these things, and Gibson obviously agreed. Thad guessed it was because Gibson himself had once been quite adventurous; he'd told Thad that during college he'd been a river guide in the Grand Canyon, and he was also an accomplished hiker.

In Thad's mind, Dr. Gibson had really taken a shine to him. And truthfully, Thad had become a little obsessed with the famous scientist. He'd read up on the man, fascinated by Gibson's work on the ALH meteorite—something that had really begun and ended back in the eighties although it hadn't been published until later. At the same time—and perhaps this was just fantasy, a concoction of his mind to build himself up to his idol's level—Thad had begun to suspect that in some ways, Gibson might even have envied *him*, because Thad was still in the adventurous phase of his life. Thad liked to think that Gibson had once lived like Thad, and now he gave speeches to co-ops.

Still, Thad knew it wasn't fair for him to compare himself to these men who had already accomplished so much. He'd only been at NASA a total of three months and one week. Maybe his enthusiasm level was a little too extreme; he knew that his vivid attitude was sometimes like a force of nature. It drove him to expect things to happen more quickly than they possibly could.

He wondered if that enthusiasm was also the source of the friction that he and Sonya had begun to experience over the past couple of months, before he'd returned for his second tour at NASA. Being at home, back in his classes at the university, daily life had seemed like a distraction. He couldn't wait to get back to Houston, back to the lab and to these scientists. Home, and even Sonya, couldn't begin to compete with what he had here.

As a pair of Indian waiters began stocking the table with strange-looking desserts on flimsy magic carpet–shaped serving platters, Thad tried to convince himself that the friction with Sonya was all in his head. It certainly wasn't her fault. It was just . . . well, being away from NASA was like standing still. He had never been able to stand still.

He was glad that tomorrow Sonya was going to get a chance to see for herself what NASA meant to him. She was flying down to stay for the weekend, and Thad had promised himself that he was going to show her why this place was so amazing. He knew she would understand, and that his enthusiasm would rub off on her. Her life at home, her dreams of becoming a model—the castings, the clothes, the nightclubs she had begun to frequent with her modeling friends—would pale in comparison with the incredible magic of the JSC.

Thad smiled as he dug into a spicy dessert that put new tears into the corners of his eyes. Meanwhile, in his mind, he pictured Sonya hanging on to his waist as he piloted a massive snowmobile across an Antarctic glacial field. He could see her red-blond hair flowing behind her, her long, chiseled legs clinging to the roaring beast beneath them. And even as his tongue burned like the heart of India, he could feel the spray of ice against his cheeks.

11

There was only one word for it.

Power.

Thad stood in a shadowed patch of grass a few feet beyond the stone path that encircled the beautiful monster, craning his neck to peer up at the closest of the five conical thrusters. Each of the thrusters was twice his height in diameter, hollowed out and stacked on top of one another in a fierce pentagon that jutted out of the base of the first cylindrical fuel stage. The thrusters were, in every curve and groove of their being, symbols of pure unadulterated power.

Thad stood alone at the base of the Saturn V rocket, even though it was deep into the afternoon and there were at least three different tour groups from Space Center Houston wandering around the rocket's park. But the tourists seemed much more interested in the long, impressive fuel stages, and the nose cone—the Apollo lunar lander that the rocket carried on its tip—where the astronauts were to have traveled. To the tourists, the thrusters were ugly and utilitarian. But Thad only had to close his eyes to see them in their raw beauty, spewing huge bursts of fire and massive plumes of smoke, breathing pure, catalyzed rocket fuel like a dragon roaring through the sky.

It was the first time Thad had actually paused in front of the Saturn V, admiring its magnificence. He had walked by the rocket a few times, when he'd needed to decompress from a long day at the lab, but he'd never just stood there before, contemplating the thing itself. It didn't matter that the heat was already starting to rise off the manicured blades of grass beneath his feet. Or that the humidity was drenching his NASA polo shirt and flattening out the curls in his hair. He felt at peace, at rest—which was a fairly unusual feeling for him.

Unusual, and so compelling that he didn't hear the golf cart approaching from behind until it was already pulling to a stop on the path a few feet from where he was standing. The clack of high heels against stone was impossible to ignore, however; he turned away from the thrusters in time to see Sonya teetering toward him, a small flight bag slung precariously over one bare shoulder while she waved goodbye to the security guard who had given her a lift from the front gate. When she reached Thad, she was smiling—but he could see the shadow of annoyance in her dark, catlike eyes.

"It's nearly one hundred degrees out here. Wouldn't it have made more sense to meet at your apartment?"

She leaned forward to kiss his cheek. Without even thinking, out of reflex, Thad reached his hands out toward her flat stomach, to work his fingers under the thin material of her tank top—but she playfully pushed his hands away.

"I feel gross right now from the flight. And I don't think you need to warm your fingers out here. If you touched that rocket ship, you'd probably burn your skin right off." Thad took the bag from her, putting it over his own shoulder. He figured there would be plenty of time for the comforts of home; they had a whole weekend together. First, it was more important to him for her to see his world, his NASA—and he knew exactly where he wanted to start.

. . .

Thad wasn't sure why he was so nervous. As he stood next to his wife at the top of the amphitheater steps, waiting for the lecture to end, his palms were damp and he was shifting his weight from foot to foot. It was foolish, he told himself; he had been in the amphitheater dozens of times. And this afternoon had begun in such a secure and comforting place—his lab, just a few floors away.

Helms hadn't been around, but Sonya had still enjoyed seeing where Thad worked. She'd been duly impressed by the futuristic setting, and she'd also gotten to meet Thad's immediate boss, Dr. Agee. Agee had seemed quite taken with her—and why wouldn't he be, she was so damn striking in her high heels and tank top. Not exactly a good fit with the NASA dress code, but it certainly brightened the place up.

But the lab wasn't the main reason Thad had wanted Sonya to visit Building 31 that particular day; he had wanted to show her the NASA she couldn't learn about from a brochure. She shared his affinity for things that were real, objects that told stories—which was why he had taken her straight to the amphitheater after the lab. Now, watching as the first-year co-ops finally began to filter out of their seats and make their way up the steps, emptying the auditorium, Thad could only hope that his efforts would be worthwhile.

It was another few minutes before he finally caught sight of Dr. Gibson's closely cropped gray hair and thick glasses as the man made his way up the steps after the last of the co-ops had exited. Gibson was wearing his lab coat over a blue button-down shirt, and he looked like he'd just come from his own lab, not just given the lecture that was one of the biggest draws of each co-op's first tour. But he did have his NASA briefcase in his right hand; Thad felt a thrill watching that case bob back and forth as the older man took the amphitheater steps, because he knew exactly what was inside.

"Dr. Gibson," he called out as Gibson was about to walk past where he and his wife were standing. "I want to introduce you to someone. I've been showing my wife around the place, and I know she would get a real thrill out of meeting an Apollo scientist."

Gibson looked up. He seemed a little distracted, but when he caught sight of Sonya—the nervous smile on her bee-stung lips and the way she pushed the hair out of her eyes—he softened. She always had that effect on people.

"The pleasure is all mine," Gibson said, emanating class. "Thad is one of our more enthusiastic characters. We're all learning a lot from him."

Thad smiled, then pointed at the briefcase.

"I was really hoping that you could show Sonya one of your cool meteorites. I know it would mean a lot to her to see one up close."

Gibson glanced down at his briefcase, then back at Thad and Sonya.

"Actually, it's a couple of big moon rocks today. The meteorites are back in my lab."

"A moon rock would be great, too—" Thad started to say, but Gibson interrupted him.

"It's no problem. We can head over to my lab—but have you guys had anything to eat, because I'm famished? Maybe a quick stop for Chinese?"

Thad could see that Sonya was into the idea; pretty cool, to get a lunch invite from a guy who had been on the Apollo project.

"Sounds like a great idea," Thad responded.

Chinese food and moon rocks—just another normal afternoon at the JSC.

. . .

Forty minutes later, Gibson was leading them down a long corridor on a midlevel floor of Building 31. Sonya was right behind him, Thad a step after, and they were moving through the building at a rapid pace. Gibson and Sonya were making small talk, and it had been like that during most of the time at the Chinese restaurant. She was taking full advantage of her time with an Apollo scientist.

As they walked, Gibson described how the building used to be, when he'd first arrived at the JSC. No matter how many times Thad heard the story, he was still fascinated by thoughts of what it must've been like for Gibson as a young man, seeing those moon rocks for the very first time. And Gibson seemed to like telling the story.

He'd barely made it to Apollo 17, the last manned mission, when he turned a corner in the hallway and pointed toward an open doorway. From the outside, Gibson's lab looked similar to Thad's, and Thad couldn't help noticing that there was a cipher lock next to the door frame, just like the one outside his lab—but because the door was already open, the lock was, of course, disengaged. Gibson told Thad and Sonya to wait in the doorway, then entered the lab. Thad figured it was some sort of procedural thing; from where he was standing, he could see that there was at least one assistant or co-op already in the lab at the time, working at one of the stainless-steel countertops. Thad couldn't tell what the man was working on—but he assumed it was something having to do with actual extraterrestrial materials, not the practice rocks he and Helms were forced to use.

As Gibson disappeared deeper into the sixteen-by-twenty-foot lab, Thad felt his curiosity getting the better of him. He decided it wouldn't hurt to stick his head inside, just to see where the man was going with the briefcase. Sonya didn't even seem to notice what Thad was doing: craning his neck, Thad caught sight of Gibson at the very back corner of the rectangular room, standing in front of what looked to be a huge upright steel safe. There was a large wheel-style combination lock on the front of the safe—but Gibson wasn't at the wheel, he was leaning over the top of the safe, the briefcase still in his hands. Thad squinted, seeing that there was a piece of paper with numbers written across it taped to the top surface of the safe. He wondered if those numbers could actually be the combination of the safe—right there, taped to the damn thing itself? It seemed a pretty foolish thing to do, but then again, this was a pretty secure environment. Gibson's personal lab, where he'd probably worked for decades. If he didn't feel secure here, he wouldn't feel secure anywhere.

Thad couldn't quite make out what Gibson did to the combination wheel after he was done with the taped piece of paper on top of the safe, but a moment later the huge door to the thing swung open. Thad saw that the safe contained five drawers, separated into compartments. Gibson bent down on one knee, opened his briefcase, and began placing the contents back into the safe. When he was done, he reached into a different compartment and retrieved a palm-sized object.

Thad quickly yanked his head back out into the hallway as Gibson slammed the safe door shut. Another minute, and Gibson was back at the entrance to his lab, smiling ear to ear. He asked Sonya to hold out her hand. When she did, he placed a small glass vial in her outstretched palm.

"This is what we call a calcareous meteorite. It's the lowest-density meteor we've ever found. They usually break all the way up when they come into the atmosphere, but this little piece survived the journey."

"This is amazing," Sonya exclaimed. "This isn't from the moon, is it?"

"No," Gibson said. "Moon rocks are a little too valuable to give away. Even for scientists like me to get a lunar sample, you have to go through numerous steps. You conceive an experiment, you write a research proposal, it goes through peer review by non-NASA scientists—there's a checks-and-balances system. Because all the moon rocks we've got came from those six Apollo missions. There aren't any more, and there aren't going to be any more. It wouldn't even be legal for *me* to own a moon rock. The ones I have in my safe, I've acquired over thirty years of research proposals, and when I retire, they'll go right back to the lunar vault."

"But this meteorite?" Sonya asked.

"That's a gift to you guys."

Sonya looked like she wanted to give the man a hug. Thad felt himself swell with pride, even though he had nothing to do with the gift.

Gibson waved their gratitude away.

"It's our job to inspire young people like yourselves. That's really the point of this place. Thank you both for a lovely afternoon."

With that, Gibson stepped back into his lab, closing the door behind him. Thad listened as the cipher lock clicked shut. Then he grabbed Sonya's hand so that they could look at the meteorite together. And for that moment, as brief as it was, all of the friction between them disappeared.

I can still hear it, Rebecca. I feel it when I close my eyes—when I open my hands to the tides of air that dance around me. For years I've held on and hoped in little whispers as I lie awake in the middle of the night alone, as the leaves are dashing near the end. My treasures are still the life-giving images that dance inside me . . . of how your eyes would light up as we talked of the next adventure—of how your body would gracefully release as I brushed your hair, of how the rest of the world always disappeared when I held your hand. I still tremble in the lonely moments, when the business fades and everything around me goes quiet. That's when I hear it the clearest. That's when the chimes echo, when my heartstrings amplify the harmony of their most treasured moments. It's always with me. It wasn't just a passing tune.

12

From a distance, the scene probably looked like some sort of bizarre suicide cult: a dozen young men and women splayed out in a circular formation on the flat surface at the peak of the giant granite dome, resting, supine, against overstuffed down sleeping bags, with backpacks for pillows—and only a pair of butane-gas lanterns to battle against the encroaching soup of night.

Thad was at the center of the human circle, crouching over his own sleeping bag. He'd been frozen in that awkward position for a good few minutes, and by now most of the makeshift campsite was watching him. He smiled toward the person closest to him, a mousy girl with spiky blond hair, dressed in boy shorts and an oversized NASA tank top.

"I'm conflicted," he said, his voice little more than a whisper. "On the one hand, the geologist in me wants me on my stomach, because this is the kind of rock formation you don't see every day. On the other hand, the astronaut in me wants me on my back. Because this is like a movie theater, with the entire solar system splashed across the screen."

The girl smiled shyly back. Even that little bit of effort pasted splotches of red against her lightly freckled cheeks. She was one of the few campers whom Thad didn't know very well. Her name was Sandra, and she had signed up for the weekend excursion at the very last minute.

"I guess the astronaut has to win, right?" Thad continued, finally rolling onto his back. He cupped his hands behind his head. He could feel the hard, smooth granite beneath his fingers, but the pressure didn't bother him. The rock was even warmer than the southern Texas air, and it felt almost therapeutic against his skin.

As usual, the weekend excursion had been Thad's idea, and he had organized everything, from the rental cars that they'd used for the three-hour trip from Houston to the purchase of the camping provisions and sleeping bags they would need for the forty-eight hours in the wilderness. Well, "wilderness" was a bit of an exaggeration; Austin State Park—specifically the Enchanted Rock Nature Area—was a well-traveled camping and hiking park. From the four-mile loop that encircled the granite rock formations that gave the park its name to the numerous caves, natural streams, picnic areas, and even playgrounds that dotted the well-preserved landmark, the place was really an outdoor amusement park for camping novices. Heck, there were even showers and restrooms.

But there was also the great granite dome, where Thad and his small group were camped. Rising almost one hundred feet over the park, it was an upside-down bowl of rock, sheer in some places, jutting and rough in others. What was really cool about the dome was the steep slope toward the summit; the traction was so phenomenal you could practically walk straight up it, your boots crunching against crystal as you went. And when you reached the top—the view seemed to go on forever.

Although the stars were just starting to come out, Thad already felt like he was watching the beginning of a fireworks display. The moon was so bright it was overwhelming the butane lanterns, and Thad was having no trouble reading the constellations without even shifting the position of his head. He had spent many nights like this back in Utah, at the observatory that was his home away from home. But here, in the wilderness of the state park, on the top of a granite mountain—it was the kind of place that turned even cynics into romantics.

It wasn't exactly legal to be camping right there on the top of the dome. There were signs all over the park warning against it, but Thad had felt an obligation to his little crew of co-ops and interns. They were his charges, and he was their social director. A stupid park ranger's whimsical rules shouldn't be allowed to keep the nation's finest budding scientists away from a view like this.

"You have to admit, it's worth the price of admission."

Again, the girl just smiled. She was cute, in a very young, almost Disney cartoon way—like the pretty little mouse that would break into song at any moment. It was easy to see that she was incredibly shy, still trying to figure out the world around her. Sweet, innocent, eighteen years old, she was an intern—which meant she was probably a college freshman, lucky enough to be spending time at NASA. That she had even signed up for the camping excursion was impressive. In a social setting like this, she was completely out of her element.

"I wonder how many of us will ever have a chance to go up there," Thad continued, pointing lazily at the canopy of stars. "I guess that's why we're all at NASA, but most of us will go on to do other things."

Thad could tell that the girl was finally working up the nerve to say something in response. He waited, trying to make it easy for her.

"If anyone can make it," she finally said, and even her voice was mouselike, "I think it will be you. But I kind of hope it will be me."

Thad grinned. She had some edge to her, after all. For an engineering major.

"Even the first Apollo capsules had room for two," he joked back.

"But only one of them got to be the pilot."

"The pilot? That's the chickenshit position. It's the other guy who steps outside the capsule door. Who got to hit golf balls on the moon. It's the other guy who's willing to take risks."

"Anyone who wants to be an astronaut has to like to take risks."

Thad was surprised to see the mischief in her thin smile. Thad liked her immediately—not in a romantic sense, even though her hard little body filled out the boy shorts and tank top in a remarkable way.

But the fact that she was fighting her shyness to joke with him—it was platonically interesting. He had a lot of acquaintances at NASA, but other than Helms, nobody he would call a good friend. That was the way his life had always worked. There was Sonya, and then there was everybody else.

Recently, there was barely even Sonya. It had only been three weeks since her visit to the JSC, but things had gone downhill pretty quickly once she'd returned to Utah. That weekend had been incredible, a sort of honeymoon, inspired by the gift of the meteorite from Dr. Gibson. But the minute she'd stepped back onto the plane, her demeanor toward Thad had seemed to change. When they talked on the phone, all she wanted to speak about was her modeling, her life in Utah. It was getting harder and harder for Thad to live in both places at once.

Here, on the top of the granite dome, his face brightly lit by a canopy of stars, it was frighteningly easy to forget about Sonya.

"It's easy to talk about risk," he countered, now facing Sandra head-on. "It's a lot harder to live it."

Sandra had inched a little closer, maybe because she didn't want any of the other co-ops or interns to hear what they were saying. Thad didn't want to lead her on, but he was enjoying the attention.

"You think I'm all talk?"

"Actually, I didn't think you talked at all. I wonder what else I got wrong about you."

"I guess that's something you'll have to figure out for yourself."

Thad grinned. She was an eighteen-year-old girl trying to sound older, trying to impress him by pretending she wasn't scared. He felt it was his duty to help her along. To open her up. If she really wanted to be an astronaut, she'd have to break out of her shell.

"You think you're ready to do something risky?"

Sandra's smile tightened a little, but she seemed to fight through it.

"What, exactly, did you have in mind?"

. . .

"You sure about this?"

"I think so."

"Because if you're not sure—"

"Thad, just shut up and take off your clothes."

Thad laughed; it was hysterical, hearing that Disney-mouse voice giving him an order like that. He guessed that she was terrified by now, because even he was feeling butterflies. It was pitch-black where they were standing, beneath the thick overhang of trees—so dark he couldn't see his own hands. But even so, they were right out in the open, and only about a twenty-minute hike from the top of the granite dome where the other co-ops and interns were presumably still sleeping. It hadn't been hard to creep out of the campsite and make their way to the bank of the winding stream that ran down one of the natural ravines. But now that they were standing there, side by side, barefoot beneath the trees—it felt so deliciously wrong.

"On three," Thad said, reaching for the buttons of his shirt.

"Way ahead of you," Sandra replied.

The next thing Thad knew, a little white tank top landed on his face, making him blinder than he already was. He laughed, yanking the soft material free—just in time to catch a flash of moonlit, lightly freckled skin racing down the bank toward the gurgling stream.

He hastily went to work on his shirt, giving up on the buttons and ripping it over his head. His belt gave him a little trouble, and by the time he had his pants down around his ankles, he could hear her splashing into the water, giving off a little squeak of pleasure as she submerged as deep as the shallow stream would allow.

"And you claim you've never done this before?" Thad yelled as he ran forward, his bare feet turning against the damp mud of the bank.

His underwear came off in one motion as his left foot hit the water, and then he was diving forward with total abandon, his chest seizing as the icy water splashed against his skin.

"Holy crap, that's a lot colder than I expected!"

Sandra laughed, and he followed her voice with his eyes. She was about five feet away, kneeling down so that the water covered her all the way to the top of her chest. He did his best not to look too carefully. There were protocols to skinny-dipping, and since this was Sandra's first time, he was going to make sure no lines were crossed. The only thing he wanted out of her was friendship.

He sank low into the water, settling into a cross-legged position on the floor of the stream. The water barely reached the top of his biceps, and the gentle pull of the current felt good against his muscles.

Naked in all respects, he found it very easy to talk to Sandra. Without even meaning to, he opened up to her about everything that was going on in his life. From the problems he was beginning to have with Sonya, to the wonder he felt working at NASA, to the future he hoped to build. To his surprise, Sandra began to open up to him in return. As he had suspected, she was extremely insecure. She was a freshman at the University of New Mexico who had grown up in a very small town outside of Nashville, Tennessee. She was a straight-A student, doing well at her internship at the JSC—but she still felt out of her league around the more aggressive scientists who populated the campus.

Thad did his best to convince her that she had nothing to be insecure about. She was as smart as anyone there, and she was a step ahead of most of the girls her age. She had as much of a shot at becoming an astronaut as any of them. She needed to get over her shyness. She needed to explore the world, to collect experiences.

"I know," she said as she listened to his pep talk, her arms crossed against her naked chest. Thad could see the hint of a pink nipple beneath the crook of her elbow, but he really tried to avert his eyes. "That's why I signed up for this weekend. And that's why I've been thinking so much about the contest you started last year, before I got here."

"You heard about that?"

Actually, Thad knew that every week, co-ops were still out scrambling around NASA, trying to push the limits of their security clearances, trying to accrue experiences they could later talk about at the swimming pools spread across Clear Lake.

"Yeah, and I've already planned out what I want to do. But I'm not going to tell you. Because when I do it, I'm definitely going to win."

"I don't doubt that at all." Thad no longer participated in the contest now that he was on his second tour. But he was pleased to hear that he had inspired those younger than him.

"What's your next act?" Sandra asked. "I don't think you can skinny-dip your way into the history books."

It was a subtle challenge—from a naked eighteen-year-old in a stream in the middle of nowhere—but Thad took it seriously. As always, he hated the idea of standing still, and the past few weeks—few months—of his relationship problems with Sonya seemed to be dragging him backward, which was even worse.

The current pushed him along as he released his grip on the floor of the streambed.

"I guess I'm waiting for the next opportunity to show itself."

"You don't strike me as the kind of guy who waits for things to happen."

Thad playfully splashed water at her, and she ducked to keep it out of her eyes. The motion revealed more of her moonlit skin, but Thad nobly turned away. He lay back in the cool water, forcing his muscles to relax, willing himself to just float.

She was right, even though she didn't know him at all. He wasn't the type to just sit back and wait for something fantastic to happen.

But he was pretty certain—at NASA, you never had to wait very long.

13

A week later, Thad was sitting at the computer desk at the back of his lab, working his way through the final pages of a project involving the solar asteroid belt. The project hadn't originated in his department—but that was one of the cool things about working at NASA. There was so much freedom, a co-op with more than the usual motivation could get involved in all sorts of fascinating things.

During his first tour, he had spent a lot of his time getting to know everybody, simply by walking around Building 31, asking plenty of questions. Through those chance encounters, he'd gotten himself attached to a number of wild experiments. He had spent three weeks helping to shoot cannons loaded with different geologic materials, measuring the impact craters, trying to reverse-engineer from those craters to identify the rocks that had caused the damage. In another instance, he had spent ten days analyzing computerized models of solar flares—simply because he had struck up a conversation with a solar specialist while waiting in line at the cafeteria.

And at the moment, Thad found himself helping out with the cataloging of all the asteroids in the asteroid belt—separating them into different classes based on their geologic makeup, searching for trends that would help determine where these asteroids might have originated. It was something he had stumbled into in his spare time, and it was supposed to have taken him the rest of his second tour. But just a few weeks after starting, he'd almost finished compiling the data, and he was hoping to return it to the planetary scientist two floors below in less than a week.

Because of the free atmosphere of the JSC, Thad wasn't surprised when a woman he knew from various hallway interactions—Dr. Andrea Cooper—wandered randomly into his lab, caught sight of him at the computer—and instantly made him forget about asteroids, solar flares, and impact craters. Dr. Cooper, he knew, was a scientist involved with the Lunar Lab, one of the few places at NASA he'd still never been to. When she pointed a finger at him and gave him a pleading smile—he was immediately all ears.

"You don't look very busy. I'm in need of someone to help me with an inventory job. Next Tuesday morning—but it might take all day, maybe two. It's gonna be sheer torture, and you won't get any recognition for it. Sound good to you?"

Thad was just glad Helms wasn't there to get the offer before him. Cooper was basically inviting him to go into the Lunar Lab, and probably into the vault—the inner part of the lab, where the samples themselves were stored. He didn't care how painstaking the job might be. It was like being invited into the basement of the Smithsonian.

"I think I can free up my schedule."

The woman gave him a thumbs-up, disappearing back into the hallway.

Thad was going to get the chance to handle real moon rocks.

...

For the next five days, Thad thought of little else. By Tuesday, his anticipation was nearly unbearable, and when Cooper finally escorted him and a technician from Lockheed Martin, one of the contractors responsible for keeping the Lunar Lab running, his energy level was so high he had trouble staying a step behind the scientist. Given a chance, he would have pushed right past her and into the vault on his own.

"It's like stepping into a submarine," Cooper said as she led them into a stairwell with cement walls. She was carrying a sheaf of computer printouts, which contained the list of random lunar samples that they would have to check and inventory. The samples were cataloged by mission; samples brought back by Apollo 11 began with the number 11, followed by the catalog number—the first Apollo sample was therefore 110001. The second, 110002. And so on.

"And I don't just mean the decor," Cooper continued, taking them up the flight of stairs. "The place was designed to be an entirely self-contained, atmosphere-regulated building. It went up in 1979, a level-four construction project—which means it can withstand a category-five hurricane without any water damage."

The guy from Lockheed whistled low, though of course he'd already know all of this. Thad glanced back at him, noticing that the man was pretty bulky. He'd probably played football in college. Thad guessed he was now a mechanical engineer, assigned to NASA as part of some enormous government contract.

They reached the top of the stairwell and came to a large steel door, with a cipher lock halfway up the frame. Dr. Cooper punched a number, then flashed her NASA ID in front of a camera pointing down from above the door. The electronic lock clicked, and the door slowly swung inward. She ushered the two of them inside.

Thad found himself in a small changing room; there was a bench in the middle of the room, a couple of sinks, and a row of lockers. Next to the lockers was a clothing rack, containing a dozen white bodysuits wrapped in plastic. Dr. Cooper waved the computer printouts at the suits.

"First we put on the suits. In the lockers, you'll find white booties that go over your shoes. You'll also need gloves, a hairnet, and over that a white surgical cap."

This was going to be really fucking cool. Taking the scientist's lead, Thad and the tech carefully put the white suits on over their clothes. It took a moment to figure out how the clasps worked, but finally Thad got the thing secure over his body, elastics closed tightly around his ankles, wrists, and neck. He retrieved a pair of booties from one of the lockers, pulling them on over his shoes. Then the gloves, hairnet, and the cap.

"You guys look like a million bucks," Cooper said, finishing with her own cap.

"I feel like an ice-cream man," Thad responded. "Or a brain surgeon."

Cooper directed them toward a glass door at the rear of the changing room. She pulled the door open, pointing them into a small, cube-shaped room with glass walls and a steel-grated floor. Stepping inside, Thad noticed that the ceiling was also grated, and the place had a very claustrophobic feel; it was so small he could touch both sides with his hands. Cooper shut the door behind them.

"This is called the 'clean room.' It's designed to cycle out any dust particles that are in your lungs. Don't worry, it won't hurt a bit."

There was a sudden change in pressure, followed by a light, antiseptic-tinged breeze flowing through the room. Seconds ticked away on a digital readout affixed to one of the glass walls. After exactly one minute, there was a loud buzz—and the door on the opposite side of the clean room clicked open. Dr. Cooper gestured again, and Thad, who was closest, pushed his way through.

He found himself in a lab—but it was nothing like the lab he and Helms called home. Instead of stainless-steel countertops, there were huge Plexiglas nitrogen chambers, with sealed, valvelike hatches. Some of the hatches had rubber gloves attached to them; others were designed to support wicked-looking microscopes, as well as other devices Thad couldn't name. There were also vents on the ceiling, along with large digital readouts that showed the oxygen and nitrogen levels in the room. He guessed that the entire place could be flooded with nitrogen, if an experiment warranted it; in that case, he'd have to be wearing one of the full-scale Racal bodysuits with self-contained air that he could see hanging along one of the walls.

Cooper led them through the lab, and at the very back they came to what Thad assumed was the vault door. It wasn't hard to identify—the thing was truly massive. Huge and steel and menacing, it was like something from a nineteenth-century bank. There was a giant metal wheel in the center of the door, next to which Thad could see another complicated-looking double-wheel lock.

Cooper approached the lock and entered a five-digit number. Then the tech from Lockheed Martin took her place, twisting in his own code. Both codes were needed, it seemed, to get the thing open. When the tech was finished, Cooper pointed Thad toward the giant wheel.

Thad nearly leaped forward, putting both gloved hands on the cold metal. With some effort, he gave the wheel a spin, and listened as the locks disengaged. Cooper told him to pull, and he put all of his weight into it, leaning back so far that he was practically hanging off the thing. To his surprise, the door didn't budge.

"You sure you put in the right combo?"

"It's designed to withstand a pound of C-4," Cooper responded. "Put some backbone into it."

Thad gritted his teeth and tried again, using all of his strength. Slowly, the thing started to move. There was a hiss as the overpressurized, nitrogen-tinged air inside the vault whispered out, leveling out with the air from the outer lab. Inch by inch, the door glided outward—until it was open enough for them to slip inside.

The vault was much bigger than Thad had expected, extending deep into the building. Along the walls he saw row after row of huge aluminum cabinets. They started about two feet off the ground, running upward all the way to the ceiling. The cabinets contained shelves and compartments, each about eight inches wide, two inches tall, extending a foot and a half back into the cabinets.

Cooper crossed to the closest cabinet, reaching for one of the boxes. She gripped it in one gloved hand, gave it a pull, and Thad watched as it slid outward. There was a metal seal on the top of the box. Cooper broke the seal, and inside were numerous samples, each individually wrapped in a Teflon bag, cataloged by number.

"So you see how this works," she said. "There are about a hundred and ten thousand samples in here, varying in size from a couple of pounds all the way down to micron dust. Each one is labeled, and our job is to go through them, find the random samples from this list, open the metal seals on the sample boxes, verify that it's there, put the lid back on, and then reseal it."

"A hundred and ten thousand samples," Thad repeated.

"Yep. Eight hundred and forty-two pounds in total. I told you this was going to be fun."

Thad peered into the open cabinet in front of Cooper. He could see into the plastic bag; it looked to be the size of a small pebble, dark, mildly pockmarked. He couldn't tell for sure, but it had to be at least thirty or forty grams. He remembered what Gibson had said in his lecture—that someone had once tried to sell a single gram of moon rock for $5 million. Forty grams—well, that would mean that the little plastic bag might very well have a street value of $200 million.

It was a crazy thought—but to Thad, no amount of money could define such a treasure. And yet, glancing around that room, at all the aluminum cabinets that lined the walls—*one hundred and ten thousand samples, eight hundred and forty-two pounds* . . . it was a staggering thing to contemplate. It was like Fort Knox, except there weren't any armed guards or men in military uniforms keeping watch. There was just the shared respect of men and women who valued science—and the historical nature of the Apollo missions—more than any amount of money.

As Cooper and the tech went to work finding the samples listed on the computer printouts, Thad lingered a moment longer, looking around the room—and suddenly noticed a small door in the far corner. The door was only about three feet high, with another cipher lock on the outer edge.

"Dr. Cooper, what's in there?"

She looked up from her computer list.

"That's where we keep the return samples."

"Return samples?"

"That's right. The rocks that have been sent out, studied, and sent back."

Thad stared at the little midget door. Of course, the return samples would be kept separately—they'd been taken out of the pristine, controlled environment of the vault, used in experiments—they weren't useful as research samples anymore. But still, it seemed odd that they would be locked away in an even deeper corner of the vault.

"Don't worry," Cooper continued. "We'll be inventorying them as well. There's a safe in there, a few feet tall, it's really kind of cute. Even though the return samples themselves are basically considered trash."

Trash—that seemed like a particularly harsh way to describe the return samples. They were still moon rocks, brought back by hand by the Apollo astronauts. Thad had a strange feeling—like he was suddenly back in the museum at the University of Utah, sifting through crates of fossils in a storage basement. *One man's trash, another man's treasure.* Except, in this case, it was such an unbelievably significant treasure. It seemed shameful to think of it as trash.

"The return rocks—they're still just as valuable as the rest, right?"

"The whole point of this place is to house lunar materials to be used by scientists for experimentation. The monetary value of these rocks is kind of beside the point. And I wouldn't get hung up on the whole trash concept—only about two percent of the entire collection is in the return vault."

Two percent of eight hundred and forty-two pounds. Thad did the math in his head. That meant there were seventeen pounds of moon rock in the return vault. Locked away in a safe designated as trash. Seventeen pounds, at 453 grams to a pound—that was 7,701 grams of moon rock. At $5 million a gram . . .

"Forty billion dollars," Thad whispered, staring at the midget door.

"Forty what?" Cooper asked, distracted by the computer printouts. She had already moved on from the topic and was no longer interested in the midget door or the return samples. But Thad could think of nothing else.

It seemed like such an incredible waste. True, the Apollo samples were much more than their monetary value—they were national treasures, they were symbols of man's greatest achievement—Thad's opinion hadn't changed since the day he'd listened to Everett Gibson give his lecture in the building next door. But this little door in the back of the vault, the thought of a safe where return samples were locked away, considered trash—it was just hard for him to ignore.

Just a few feet from where he was standing, there was a little door that led to a safe containing forty billion dollars' worth of something NASA considered trash. Locked away, in the dark, where nobody could see it or touch it or even know that it existed. *Forty billion dollars.*

Thad shook his head, then finally pulled his eyes away from the midget door and joined Cooper and the tech as they began the long process of inventorying the lunar vault. But inside, his thoughts were still percolating. He had no idea where this mental process was going to lead him, but the knowledge he had just obtained wasn't going to simply disappear. His mind didn't work that way.

Unlike the scientists at NASA, he couldn't simply file away what he'd just learned was behind a little door, in a safe marked *trash*.

14

Thad didn't sleep well that night. By two in the morning, he was tossing and turning in his dorm room–style twin bed, trying to find a position that might just be comfortable enough to make his mind shut down. He wanted to believe that his inability to fall asleep was due to the ache in his muscles; the inventory job had been just as torturous as Cooper had warned, a full fourteen hours spent bending and sometimes crawling as they worked their way through the printouts of random samples. And that heavy vault door—Thad had strained muscles all across his back working it shut after they were done.

But Thad knew it wasn't the physical labor that had set his thoughts on edge. Every time he closed his eyes, he saw that miniature door—and what was inside. At one point during the inventory work, Cooper had led him through that little door to the safe. Inside that safe, those lunar samples looked exactly the same as any of the other rocks, because, to the naked eye, they were exactly the same. Maybe they had been exposed to air, dipped in liquids, maybe some scientist somewhere had applied pressure or heat—but that didn't change what they were. Apollo moon rocks, national treasures, infinitely valuable.

Thad didn't know why the thought was driving him so insane; hell, he'd been around valuable materials before. At NASA, every piece of equipment he worked with in the lab was worth more than his car, more than everything he had in his bank account. Money had never been that big a deal to him before, mainly because he'd never had any. He and Sonya were in debt, had always been in debt. At the moment, Thad owed about $6,000 on his credit cards, maybe another five in school loans. But it couldn't be about that—how was today different from a week ago, or six months ago, or two years ago?

But today *was* different. His relationship with Sonya was in trouble, maybe even on its way toward ending. And here at NASA, he'd become this adventurous, impressive character—everyone knew who he was, everyone wanted to be around him. And yet he knew, deep down, that it was partially an act. It was a reinvention, because deep down he was this shy, messed-up kid who'd been kicked out of his house, who'd gotten married too young, who wanted to be an astronaut but probably would never have the chance.

Lying in his bed, in the middle of the night, it was the first time he'd really let the truth resonate inside of him. His chance of becoming an astronaut, of becoming a piece of human history like Everett Gibson—it was beyond improbable. The odds just weren't in his favor. He had no connections, no fallback plan, no means to compete with the kids who could pay their way through life.

And yet he was living each day as if it was just a matter of time. He was swept up in the fantasy of being the first man on Mars—the same way he'd created the fantasy of who he now was, this adventurer, this James Bond type of character who could do anything, who would do anything. Fantasy was his true talent. Fantasy had always been his true talent, the cloak he'd wrapped himself in to protect him from the things he couldn't control.

And now, suddenly, his fantasies had been given a new element to work with, and for reasons he could not entirely explain, that element was pushing him in a direction—just a nudge, at first, nothing more than a nudge. But it was palpable, nonetheless.

15

Gray on gray on gray. Thick and dark and ominous, like the intertwining ropes of an immense fishing net cast across the sky, swallowing up every inch of visible air, obscuring everything, even the muted glow of the nearly full moon.

Looking up at that angry sky, Thad knew he was about to get soaked. Still, he remained right where he was, flat on his back on the cool cement roof of the South Physics Building of the University of Utah, his head resting against a faded couch cushion as he watched the clouds block out the moon. It was a little after ten, and he had been lying like that on the roof of the building for a good two hours; he'd already ignored a half-dozen calls from Sonya on his cell phone, letting the annoying, electro-pop ring tone he had selected to represent the girl she was rapidly becoming reverberate off the walls of the domed observatory that rose up on the rooftop behind him. There was a time when Sonya would have accompanied him to the weekly Wednesday-night Star Party, which he had originated when he had resuscitated the comatose observatory—but tonight Sonya, as she had done the last few Wednesdays, had joined her model friends at a downtown dance club, leaving Thad alone with the gray-on-gray sky.

Of course, there wouldn't be any Star Party tonight. The clouds made sure of that. Nevertheless, Thad's duty to the observatory—which he had personally rebuilt, lobbying the university for the funds and equipment, turning astronomy into one of the most thriving extracurriculars on campus—was still a good excuse to avoid a night wasted in some noisy, smoky club; no matter how bad the weather, there were always a few stragglers who would show up. It didn't always dawn on people: if you couldn't see the stars through the clouds with your naked eye, a Celestron eleven-inch mounted telescope wasn't going to make any difference.

Although Thad been the one responsible for the growing popularity of the astronomy club, the structure of the observatory had been around for nearly fifty years. Built on the roof of the South Physics Building in 1976, it had been through a number of changes in the past few decades. Recent additions of a pair of high-tech telescopes, a few cameras, a spectrograph, and some shiny new mounts and housings had turned the place into a first-class stargazing facility. As soon as Thad had returned from his second tour at NASA, he'd gotten the Wednesday-night Star Parties going, and over the past few months they'd grown from a handful of moon-obsessed telescope freaks to a real social gathering, sometimes numbering in the dozens.

Thad was proud that he'd been able to bring aspects of his reinvented personality back to the university with him. But sadly, his new persona hadn't helped him at all with his relationship problems. Sonya had also become more outgoing and social, but her new friends, and the kinds of places they liked to go—Thad didn't have anything in common with her world at all. So he had simply stopped accompanying her to the casting calls, cocktail parties, and especially the nights out in the dance clubs.

Lying outside on the roof of the university building, the observatory rising up behind his back, staring out into that gray-on-gray sky—here he could pretend he was back at NASA. Back in a world of science and fantasy.

Thad's thoughts of NASA dissipated as the sound of a door opening and closing echoed across the desolate rooftop. He didn't lift his head as he heard the approaching footsteps. From the way the stranger's boots shuffled against the cement roof, he could tell that the person was either drunk or on his way to being so, which meant he was probably one of the Star Party regulars, either too soused or too stupid to realize that a cloudy sky looked like a cloudy sky, even through the most powerful telescope on Earth.

"Thanks for coming," Thad said, without lifting his head from the couch cushion, "but this week's Star Party is canceled, due to the lack of stars. Come back next week, and hopefully we'll have something to look at."

The steps didn't even pause, shuffling closer until they were just a few yards away. Thad heard a grunt as someone lowered himself next to him on the roof, and then there was the sound of a lighter flicking on and off. A sickly-sweet, decidedly herbal puff of smoke floated past Thad's face.

Thad had nothing against marijuana use, although he didn't touch the stuff himself, but he was surprised that someone would smoke the illegal substance right out in the open, on the top of a university building. Any moment, a professor or a security guard could wander out—and in fact, often the astronomy TAs came for the Star Parties, although an astronomy TA would know better than to come up to an observatory on a cloudy night.

Curious, Thad raised his head to look at the visitor. The guy was sitting cross-legged, his back against the wall of the observatory, his arms crossed against his chest. He was wearing a wool cap pulled down low above his eyes. His jeans had holes in them, and he was wearing gloves with the fingers cut out. He looked kind of homeless, but where his face was lit up by the tiny marijuana cigarette, Thad could see that he was young, maybe even younger than himself. There were patches of facial hair on his jaw, and his cheeks had the ruddy complexion of a guy who spent a lot of time outdoors. Ringlets of brown hair stuck out from beneath the lip of his cap. He was staring past Thad, past the edge of the rooftop, at the rolling view of the south corner of the university's campus.

"No worries, man," the stranger said. "Happy to look at the clouds if I can't see the stars."

There was a hint of California surfer dude in the way the guy talked. Maybe it was the weed, but he seemed so relaxed, so completely devoid of tension. Thad didn't think he'd ever felt the way that man looked. He'd always been so much more tightly wound, so infused with the energy that often came out as enthusiasm. He couldn't imagine this guy ever being described as enthusiastic.

"I guess that's a pretty good attitude."

The guy took another hit off his joint, then let his head rest back against the observatory wall behind him.

"How big you think it is? I mean, like, does it really go on forever, like they say in the books? Because how the hell can something go on forever?"

Thad assumed the guy was talking about the sky or, more specifically, the universe. Compared with the kind of conversations Thad had gotten used to at NASA, it was pretty basic and a little juvenile—but it was science, and Thad liked nothing more than to talk science. Compared with a conversation about fashion or modeling at some neon-lit club, this was as close to the JSC as he was going to get. So he lay back against the cushion and started to talk.

As they conversed, Thad learned a lot about the laid-back kid. His name was Gordon, and he was also a student at the university. He had taken a little time off here and there, but now he seemed to be on track with his studies, trying to make it through the spring semester without getting lost somewhere along the way. What Thad liked most about Gordon was that he seemed extremely curious about the big questions in life. About the size of the universe, about how many stars there really were, about the possibilities of life on other planets. Surprisingly, at the same time, Gordon was very religious, and it became clear almost immediately that he had grown up in a Mormon environment very similar to Thad's. Halfway into the conversation, Gordon mentioned something about losing a wife and kid to the Mormon Church, which Thad didn't press him on; crazy, that at their ages they had both already been married, but that was Utah. Somehow, Gordon had remained close to his mother and uncle, and had also retained much of the Mormon teachings. Interspersed with his questions about science, he often quoted Mormon Scripture. The possibility of life on other planets seemed to conflict him, but that didn't keep him from trying to dig deep into the idea.

Thad liked the guy, and also thought there was something very bold and exciting about him; he had the balls to just sit out there on the roof smoking pot, talking about aliens—yet deep down he was still this well-mannered Mormon kid. Adding to the conflict in his personality, somewhere in the conversation Gordon mentioned that he had a bit of a criminal record, something small and insignificant, but there nonetheless. Things had been bad for him for a while, but now he was back at school and he was doing well. He wanted to hang out with people who were going to be good for him—and he really liked hearing that Thad was a triple major—obviously someone who worked hard and was moving in the right direction.

By the time midnight rolled around, the half-dozen calls from Sonya had become more than ten, and Thad knew it was time to get moving. He and Gordon agreed to keep in touch, whatever that might mean. Not a formal thing—just an agreement that if they ran into each other on the street, they would maybe meet for lunch.

As they headed toward the elevator that would take them off the roof, a light rain began to fall; Gordon didn't seem to notice, maybe because he was on to his second joint by then, or maybe because his hat was pulled down so low he couldn't feel the drops. But Thad was shivering as the dampness worked its way into his bones. Something about

Gordon had inspired a thought: the two of them might never become friends—but that didn't mean they wouldn't have an impact on each other's life.

Sometimes, Thad knew, as a scientist, it was the molecules that only briefly touched that caused the biggest reactions.

. . .

Thad ran into Gordon a handful of times over the next few weeks. Twice at one of the campus dining halls, when Thad was just on his way out and Gordon on his way in. Once, crossing in front of the geology building, Thad walking with a couple of girls from one of his science classes and Gordon just sitting there, on a bench, smoking another joint. And then a third time, on the steps of the main library. Gordon was drinking something out of a thermos, still wearing his wool hat and cutoff gloves, and Thad paused in front of him, coming to a sudden decision.

Thad hadn't intended on bringing up the topic—but seeing Gordon on the library steps, right in the bright light of day, he decided maybe it was time to put words around the thoughts streaming through his head. Throw it out there, see where it landed. But not here, in the middle of the campus.

"You doing anything right now?" he asked.

Gordon looked at him like he was crazy.

"Solving world hunger. Why?"

Thad grinned, and beckoned Gordon to follow him.

. . .

Twenty minutes later, Gordon was still following as Thad strolled through a deserted wing of the University of Utah's Museum of Natural History mineral collection, pointing out the various crystals and unique fossils that lined the lit-up glass cabinets that ran along the walls. It was a pretty good collection, and Thad had spent numerous hours wandering through the museum, alone and with Sonya. At various points in his time at Utah, he had worked cataloging these very minerals and fossils, carting them up and down from the basement storage areas of the museum—and yes, he had borrowed a few to display in his apartment, but nothing as valuable as the specimens in front of them now, specimens like the imprints of a brontosaurus foot off to their left, or the shiny green jade deposit encased in brightly lit cubes straight ahead.

Thad was surprised to hear that Gordon had been in the museum before as well; in fact, just the summer before, Gordon had donated a rare angel wing calcite crystal to the university, which was now tagged with the note Gordon had written himself: *In Memory of Kelen McWhorter*—a sister who had died in a car accident five years earlier. Thad had heard of the crystal, and he told Gordon that he believed it was worth a fair amount of money. Which kind of led, naturally, into what Thad had brought Gordon to the museum to discuss. And that's all it would be, in Thad's mind, two college kids having a conversation. Nothing anywhere near as dangerous or illegal as lighting up a joint. Just a conversation, words, the expression of a little fantasy that had been building in Thad's mind.

"I've been thinking about something," Thad started, "and I wanted to get your opinion."

"People don't usually come to me for advice," Gordon responded, peering into a case that contained fossilized insects from the Jurassic period.

"Okay, not advice, really. I'm just trying to figure something out. See, I think I might be able to get my hands on something valuable."

"Like dinosaur fossils?" Gordon pointed, grinning. Thad laughed.

"Even more valuable. And I'm trying to figure out if maybe it would be possible to find a buyer. Like, on the Internet or something."

"World is made up of buyers and sellers. You got something to sell, there's usually someone out there who's willing to buy. And the Internet—you can find a buyer on the Internet for anything, if you look hard enough."

It was the kind of answer Thad had expected from Gordon. Thad's thinking was, since Gordon obviously knew something about drugs, and had already pointed out that he had a little criminal record, maybe he knew people in the underworld. Maybe he had some weird connection into some underground market somewhere.

"So if I did have something valuable, you think you'd be able to find someone to buy it?"

"Depends what we're talking about."

Thad swallowed. Was he really going to say this out loud?

"Moon rocks."

Gordon looked at him, then started to laugh.

"You know that's all bullshit, right?"

"What's bullshit?"

"The whole moon-landing thing, man. Couple of guys, walking around hitting golf balls, planting some freakin' flag—you think any of that was real?"

Thad rubbed his eyes, trying to figure out if the stoner was playing with him, or actually meant what he was saying. He realized that he had never explicitly told Gordon that he worked at NASA. Gordon had no idea that Thad hung around men who had actually been there when Neil Armstrong had walked on the moon.

"So you think—" Thad started, and then stopped himself. He wasn't going to get into an argument about conspiracy theories. "Look, here's the thing. I might know somebody who can get his hands on a couple of rocks. They're worth a lot of money, if we can find someone who's interested. A collector, a gem dealer, someone like that."

Gordon was tapping his fingers against a glass fossil case.

"Moon rocks. From a museum? A private collection?"

Thad shrugged. He didn't want to go into it any deeper than that. The kid could believe whatever he wanted. That Thad knew about a moon rock locked away in a basement drawer in the same museum they were walking through. Or that some member of the royal family of some South American country had a rock he wanted to fence. Who cared? Thad just wanted to know if Gordon could help him figure out if there was any way to sell a moon rock, if he had one.

"I could probably figure that out," Gordon finally said. "Do a little research, send out some e-mails. I'm pretty good with the Internet."

Thad nodded, his excitement rising. Was there really anything wrong with Gordon sending out a few e-mails? Would it really be a big deal? Thad hadn't done anything wrong yet—and he was probably never going to. Hell, he didn't even think the moon-rock thing was actually possible. Just thinking about what he would have to pull off, to get inside that vault—no, it was just a mental game. Another fantasy that he was beginning to construct. That was his true talent, fantasy. He had reinvented himself as a social star at NASA, he was impressing everyone there with his adventures and his contests, his enthusiasm—all of it. These thoughts, they were just a natural progression—another adventure, but this one so far-fetched it would likely remain lodged in fantasy.

Thad continued on through the museum, moving from the fossils and minerals to an area filled with precious mosaics that the museum had borrowed from a collector in Turkey. By the time they reached the building's exit, Gordon had his wool hat pulled down low over his eyes, the thermos out from under his coat. For all Thad knew, Gordon had already forgotten about the exchange. The moon rocks had gone back into the dark, quiet space in the back of Thad's mind.

And maybe that would be for the best.

Christ, yeah, that would definitely be for the best.

16

It was a little less than two weeks later that Thad ran into Gordon again, passing by the same library steps; Thad was moving fast, a full load of oversized physics texts cradled in his arms, barely paying attention to his surroundings because he was already ten minutes late for a lecture on quantum mechanics. Moon rocks were the furthest thing from his thoughts; they had been replaced by quarks, neutrinos, and a dozen poorly understood little particles, spinning and twisting through imaginary orbits behind his eyes. Deep into the spring semester, he was beginning to find physics almost as interesting as astronomy.

Even as enveloped in the upcoming lecture as he was, Thad almost dropped his textbooks when he heard the familiar California-stoner voice from the direction of the library steps.

"Train keeps a-rolling, eh, man? You're liable to run somebody down, moving that fast."

Thad restabilized the books against his biceps, then looked up, seeing Gordon strolling down the steps in his direction.

"Just heading to a lecture."

"Right on, brother. You and me both. But when you get a chance, you might want to check your e-mails."

Thad felt the weight of the textbooks digging into the skin of his arms. He realized with a start that he hadn't looked at his e-mails in a day, maybe longer. For the first twenty-four hours after his last conversation with Gordon, he'd checked his e-mail account every couple of hours, but in the past few days he had been looking at it less and less often. He had assumed he'd been right, that Gordon had forgotten about his request. But from the grin on Gordon's face, it was obvious now that he had been wrong.

"I'll go check it out right now."

Gordon didn't stop moving; he just crossed right in front of Thad and gave him a little wink.

"What about your lecture?"

But Thad had already changed directions and was hurrying up the library steps.

Quantum physics could wait.

. . .

It didn't take long for Thad to find an open computer terminal in a fairly isolated stall near the back of the library's 1960s-era research room. The computer wasn't anywhere near as up-to-date as the one he was used to at NASA, but it was perfectly functional, and more important, the cubicle had high enough walls to obscure the screen from any prying, nearby eyes.

Thad knew he was being paranoid as he hunched over the computer, hitting the keys rapidly, opening his e-mail account. Of course, nobody was going to be the least bit interested in what he was doing. And really, he wasn't doing anything at all, just checking an e-mail from a friend.

It took less than a minute for Thad to locate the e-mail: the address from which it had been sent was bizarre enough that it could only have come from one person.

Fractalysed@yahoo.com.

And the e-mail itself seemed as disjointed as the address. It was more than a page long, and it was obvious from the start that Gordon had cut and pasted a number of different messages together. Thad counted at least seven addresses within the body of the e-mail, all people that Gordon had either contacted or received some sort of response from.

Some of the missives seemed promising, but none were concrete. It looked as though Gordon had been shooting out almost random feelers into the electronic wasteland—beginning with a private international mineral collectors Web site located in Iraq, from which he'd managed to cull fifty or so e-mail addresses from potential collectors. Using these e-mails as his targets, Gordon had then concocted a short form letter, basically spam, which he'd mass e-mailed to the addresses. The form letter was pretty foolish sounding—especially the fake name Gordon had chosen for himself—but he had managed to get the information mostly correct. It was wild, seeing the spam letter; the idea that it was out in the open, bouncing around the Internet—it was pretty terrifying. But it was also exhilarating; although many of the responses were simply short, sometimes profanity-laced messages explaining that the sale of moon rocks was, indeed, illegal, a handful seemed to be interested.

Thad realized that he'd have to take over from here; Gordon had done his job, had made a few contacts—but Thad was the one who actually knew what they were trying to sell. If, indeed, they were really trying to sell anything at all.

Thad had spent enough time on the NASA computers to know how to set up a dummy e-mail account. As ridiculous as Gordon's chosen pseudonym sounded, Thad was forced to adopt the new name.

Using the handle, he went to work on a new missive: rereading Gordon's e-mail, he discovered that a number of the potential targets were related to a certain international Web site for mineral collectors—a sort of club for "rock hounds." Since the site was in Europe, Thad didn't feel nervous crafting an advertisement to put on the mineral club's online newsletter. He had to choose his words carefully—but he knew that the advertisement would reach the entire club at once. If these people took their hobby seriously enough to spend time on a Web site dedicated to rocks, there was a good chance that at least one of them would be interested in what Thad was purporting to sell.

As Thad drafted the ad, he tried to picture the sort of person who might respond to an offer of a chance to buy the most valuable substance on Earth. He knew a lot of people found moon rocks fascinating, but it would have to be a special sort of individual, someone desperate to actually hold a moon rock in the palm of his hand.

He was searching for a true rock hound. Someone who took his hobby seriously enough that he'd read the advertisement and immediately get that burst of adrenaline, that rush that could only come from a true addiction.

Grinning at the thought, Thad hit more keys on the computer. Even though it was little more than a game, for the moment, it was still quite possible that Gordon's e-mails, and his own crafty advertisement, were about to make some lucky rock hound's day.

17

Antwerp, Belgium

> Greetings.
>
> My name is Orb Robinson from Tampa, FL. I have in my possession a rare, multicarat moon rock I am trying to find a buyer for. The laws surrounding this type of exchange are known, so I will be straightforward and nonchalant about wanting to find a private buyer. If you, or someone you know, would be interested in such an exchange, please let me know. Thank you.
>
> Orb Robinson.

Axel Emmermann watched the green-yellow glow of his computer screen dance around the curvature of his pilsner as he expertly turned the tall glass in front of his eyes. He could even make out a few words from the strange e-mail that had arrived in his in-box just moments ago, but for the moment, his attention was more focused on the contents of the pilsner glass than its surface. The beer was incredibly light, its carefully cultivated golden coloring so profound, it could almost be described as a texture. The deep hue completely overwhelmed the handful of minuscule air bubbles that signified its gentle carbonation. Axel slowly pressed the glass to his lips, taking a small sip, letting the bittersweet, smoky mixture play across his taste buds. He noticed, with no small satisfaction, that the temperature of the beer was just about right, and the pilsner glass had allowed it to breathe well enough to satisfy his practiced palate. Thoroughly pleased, he brought the glass back to his lips and took a deeper drink.

Axel knew that if his wife—or one of his two children, aged twelve and fifteen—had wandered into the first-floor living room and caught sight of him performing the ritual of the pilsner glass and the amber beer, there was no doubt it would have been the cause of much amusement. In Axel's world, even something as simple and mundane as enjoying a late-night beer had its procedures. *Everything in its place, everything in its way*. A continually examined life, Axel liked to remark, and he really couldn't help himself: he liked things in his world to behave the way they were supposed to, whether that referred to beer kept at a precise temperature and aerated for exactly the right amount of time, or to the bigger political issues that often seemed so foreign in a place as wonderfully sedate as this quiet corner of suburban Antwerp.

Axel considered himself a self-taught Renaissance man, and he had been collecting knowledge about the way the world was supposed to work for almost fifty years now. His wife and kids liked to say he was a student of everything—which, they often remarked, was a class you could never finish. Axel knew they were probably right, that there was no end point to knowledge for knowledge's sake. But that's what made being a student of everything so much more interesting—every day there was a new puzzle you had to try to solve, which only led to the next puzzle, and on and on.

Axel drained the last drops of the wonderful beer and placed the glass carefully back on its coaster, situated near the corner of the small oak desk he had inherited from his father, years ago. To an outside eye, the small office area he had carved out of the corner of his living room might have seemed cluttered; to Axel, everything in the area made perfect sense, from the high stack of manila folders containing spectography data that covered nearly every inch of the compact bureau overlooking his garden via a small, shuttered window, to the loaded-down, handmade wooden shelves that lined the walls, filled to near collapse with cardboard boxes and sealed Tupperware containers. *Everything in its place, everything in its way.*

Except, a little after midnight in the middle of the week as a subtle rain sputtered against the thick glass of the windowpanes above the bureau, there were at least two things that did not seem to be in their place, or way, at all; Axel was awake, for one, which was easily explained, the result of a particularly heavy dinner of *vlaamse stoofkarbonaden*, a Flemish stew made with beer—though in this case not anywhere near as satisfactory a vintage as the amber concoction he had just drained. But the second puzzle seemed much more complex, and Axel knew he would not be joining his wife upstairs until he had at least begun to make sense of it.

Axel leaned forward so that his wire-rim eyeglasses were only a few inches from the computer screen, and reread the e-mail again, mulling over each and every word, like some sort of college professor studying an important and archaic text. Of course, he wasn't a professor, though his appearance could easily give off that impression: balding, with a rapidly whitening, meticulously trimmed beard, ruddy, rounded cheeks, and sometimes, especially late at night, a spiderweb of fault-line cracks at the corners of his eyes, the result of spending far too much time looking at things that were very, very small.

> Greetings.
>
> My name is Orb Robinson from Tampa, FL. I have in my possession a rare, multicarat moon rock . . .

Axel couldn't deny the sudden flush he felt in his cheeks as he reread the sentence. Like everyone else in the modern world, he got a fair amount of junk mail, spam, garbage sent to his e-mail address every day, but there was no doubt in his mind that this specific e-mail had been sent to him, specifically, because it would cause just such an excited reaction. He could guess exactly where this "Orb Robinson" had gotten his e-mail address—from the Web page of the Antwerp Mineral Club, where Axel was listed as one of the charter board members.

He was a rock hound. More than that, his obsession with rocks and minerals had been a centerpiece of his life for many years now. He still remembered where it all started: he had been around eight years old and had heard on the radio about a British Petroleum promotion in which they were giving away little boxes of Brazilian minerals with every purchase of gasoline. At the time, only one member of his extended family had owned a car, and it had taken a while to convince his uncle to drive him to the nearest BP gas station, far out on the highway that connected Antwerp to Amsterdam. But the minute Axel had held that box of rocks in his eight-year-old hands, he had become hooked. Twenty-four little pieces of rubble, ugly as hell—and yet they seemed completely magical to Axel, how each one told a story about a time and place, how each one hinted at an orderly and understandable historical record.

By age sixteen, Axel had found and joined the Antwerp Mineral Club, of which he was still a member thirty-four years later. Over the years, he had always maintained a hobbyist's interest in the things that helped him understand the rocks he collected through the club: chemistry, geology, even a little space science. But after his mandatory army stint, he had begun to find girls and beer a little more exciting. A seven-year job as a DJ in a local disco had made him just successful enough to grow his collection to about nine hundred specimens, none of them very valuable but, on the whole, quite respectable. He had also managed to collect a wife, serendipitously named Christel. It had been Christel's idea for him to reconnect with the mineral club in a more regular fashion, and by the mid-eighties he was meeting with them every Monday, helping to host visiting rock collectors and geologists from all over the world. Antwerp wasn't exactly a major stop on any European tour, but it was close enough to Paris and Amsterdam to bring in a few dozen notables over the years. Axel was quite proud of what he and the club had accomplished. Three or four of the club's more prominent members were actual professors with minerals named after them. And in fact, one of the most prominent members, Professor René Venassle, had been called to the royal palace to give a technical speech when U.S. president Richard Nixon presented an actual lunar sample to the king himself.

Rereading the e-mail, pausing on the words *moon rock*, Axel remembered that episode: how the U.S. ambassador had personally dropped off the heavily guarded sample, how it was displayed by Axel's very own club at that week's mineral show. Although at the time Axel wasn't spending nearly as much time at the mineral club as he did now, he had done a fair amount of reading about lunar samples, to better educate himself about Nixon's gift. He knew that the rocks were illegal to own, and that they were also very valuable.

Two facts that made the e-mail immediately suspect, but also a little bit exciting.

On a first reading, Axel had even wondered if the e-mailer was referring to lunar samples at all. His first thought was that the seller was actually referring to "moonstone," which was a variety of feldspar, an uncommon but not incredibly rare type of gem. Of course, the trading of moonstones was perfectly legal. Which made the next line of the e-mail make a lot less sense:

> The laws surrounding this type of exchange are known, so I will be straightforward and nonchalant about wanting to find a private buyer.

Which meant that the e-mailer was, indeed, talking about lunar samples. Axel's next instinct was just to delete the e-mail, without another thought—because if this person was talking about moon rocks, the e-mail was obviously a hoax. Axel had seen how well guarded the moon rock sample that had been given to the king was; no doubt lunar samples everywhere were kept under the same kind of lock and key. If owning a moon rock was illegal, certainly selling one would be even more so. Axel had even reached for the delete key—before stopping himself, more thoughts running through his head.

He realized that it was probably—almost definitely—a hoax. But even so, there was something odd and disconcerting about the message. Axel knew that Christel would think he was simply being his obsessive, inquisitive self, but something in this didn't seem right.

If you were going to try to sell a fake moon rock, wouldn't a rock hound such as himself be the last person you would e-mail? Because certainly, it wouldn't take a man like Axel Emmermann very long to recognize a fake; a few questions, a photograph, or even a moment eye to eye with the item in question, and he'd know its true nature.

No, you'd have to be ridiculously stupid to attempt to sell fake moon rocks to a charter board member of the Antwerp Mineral Club. But rereading the message yet another time, Axel didn't see any signs of idiocy. The syntax was good, even the man's name had a certain panache to it: Orb Robinson, like some sort of transposition of Roy Orbison—the dead rhythm-and-blues singer, who, in point of fact, had a degree in geology. It occurred to Axel that the person who wrote this e-mail wasn't stupid at all.

Which meant that quite possibly he really was trying to sell a "multicarat moon rock."

Axel was beginning to lose track of the time as he contemplated the curious e-mail, and he realized that he'd never get to sleep if he didn't investigate further. He knew enough about computers to do a pretty simple search of the originator of the e-mail. To his surprise, he quickly found that "Orb Robinson" had posted his inquiry to a fairly large number of mineral-club bulletin boards across Europe. But even more surprising, Axel found something that caused him to sit straight up in his chair.

Back on March 9, Orb Robinson had posted an advertisement on the Antwerp Mineral Club's main Web site. Axel quickly followed the link, and there it was, ad number 1275, posted to the Web site's "Virtual Quarry":

> Priceless Moon Rocks Now Available!!!
>
> "Orb Robinson" orb_robinson@hotmail.com
>
> If you have an interest in purchasing a rare and historically significant piece of the moon, and would like more information, then please contact me by e-mail and leave your contact information and an explanation of your interest.
>
> Sincerely, Orb

Axel whistled low to himself, then removed his glasses and rubbed them against his sleeve. He glanced at the empty pilsner glass, wishing he still had some of the good amber stuff within arm's reach, because now he felt like he needed a drink. He was suspicious before, but seeing the ad right there on his own club's Web site, he was beginning to be convinced. Hoax or not, he was witnessing a crime in progress.

Even though he was an avid collector, the thought of actually attempting to buy what this person was purporting to sell never crossed Axel's mind. Just as he thought that the world needed to have an order about it, everything in its place, he was a firm believer in right and wrong, that there were lines you couldn't cross, shortcuts you couldn't take. He had served in the military because it was expected and also because it was the right thing to do. He wasn't a rich man by any means, but he lived a good life in his little corner of Belgium, he had a wife who loved him and two kids who didn't hate him. And he had his hobbies, his puzzles; that, to him, was what this crime really was—a puzzle that he now needed to solve. *Everything in its place, everything in its way.*

The only question that remained was how, exactly, Axel was going to put this Orb Robinson into his proper place.

. . .

The weekly meeting of the Antwerp Mineral Club was already in full swing as Axel strolled determinedly through the wide, high-ceilinged dining hall of the Antwerp youth center. Even from the back of the hall, he could see that most of the regulars were there, gathered around the dozen or so industrial-looking rectangular tables that took up much of the center of the room. The slide projector had already been turned on, but not advanced to the first slide; the big screen that took up the entire far wall of the cantina glowed a bright, almost solar shade of yellow, backlighting the tables and the conglomeration of middle-aged, mostly bearded men who migrated around them—a herd inspecting a familiar water hole.

It wasn't the tables themselves that the herd found interesting. On top of the tables were various-sized boxes, ranging from the cardboard variety to more high-tech compartmentalized plastic cases that kind of looked like fishing-tackle containers. It was a weekly ritual; before the slideshow presentation, there was an hour set aside for the buying and selling of specimens. As Axel reached the first table, and reflexively glanced into the nearest set of boxes, he could see that it was the usual fare: shiny pieces of quartz, a few volcanic specimens, a handful of minor gemstones—nothing of particular value anywhere but here. To these men, who would offer up a perfectly good Monday night to spend gathered in a youth-center dining hall—a bit of quartz or a volcanic rock could sometimes seem like a treasure.

Axel was disappointed in himself for being late, because it certainly wasn't like him to be late. But tonight, just as it had done for the past two nights, dinner with his wife had led to a discussion about the strange e-mail—which he was now convinced was a window into some sort of crime in the making. Not an argument, exactly, because he and Christel never argued. But certainly a hashing out of opinions.

Over the past forty-eight hours, Axel had become convinced that he had to do something. But Christel, for her part, didn't like the idea of him sticking his nose into something that might end up being dangerous. If this *was* some sort of hoax, then the danger would be minor. But if somehow this person was selling a real moon rock—Axel might be getting himself involved with a dangerous character.

Axel had explained to his wife that it wasn't in his nature to just sit back and watch a crime in progress. What sort of person could stand by and see something they knew was wrong, and do nothing about it? But his wife wasn't buying any of it; she had responded by saying that despite his noble explanation, the real reason he wanted to get involved was that he thought it would be fun. Another entertainment, another hobby. Like rock collecting itself. Or "popinjay," another of Axel's passions—a strange little archery game that involved shooting a wooden bird off the top of a ninety-five-foot pole. Sometimes with a crossbow. In front of an audience.

Axel knew there was some truth in what she was saying, but he didn't want to give her the satisfaction of winning the discussion. It might be a puzzle he intended to solve, but damn it, solving the puzzle would redress a wrong.

"Mr. Emmermann," one of the bearded men hovering over a nearby tackle box filled with gemstones exclaimed, causing most of the men around him to look up and smile. "We were worried you might have fallen into the river. Or, at the very least, gotten your foot stuck in a pilsner glass on the way out of your house."

Axel grinned at the man, then made a big show of bending over the rim of the tackle box, peering at the contents.

"Actually, we're having a little issue with the foundation of our chimney. But I knew I could count on you to bring me a little worthless rubble to fill the gaps."

The bearded man feigned indignation, placing a hand over his heart. He was the club secretary—in his mid-sixties, among the oldest members of the mineral group; a postman by day, he was one of the most respected rock hounds in Antwerp—and he also knew how to run the slide projector.

"But why would you have come all the way here to buy my rubble when we all know you could have easily used your wife's stew. One spoonful in between the bricks, and your chimney would have lasted a hundred years."

Axel laughed, because he couldn't argue with the man's point. Before he could think of something witty in response, a bushy-haired, portly amateur geologist shouted over from a table to the left.

"Or maybe he's late because he was busy buying moon rocks."

Axel's ears perked up as he stood frozen over the tackle box of mildly precious gemstones. He looked toward the bushy-haired, portly man. Alfred Schnermeyer was one of the handful of Ph.D.s in the group, and for the past three years he had been the editor of the club's newsletter. Axel looked from him to the other rock hounds nearby and saw that they were all smiling, as if in on the same joke.

"Don't look so surprised," the club secretary exclaimed, giving Axel's shoulder a squeeze. "We were just discussing it before you got here."

"You all got the e-mail as well?"

"Everyone on the club's main Web page. This Orb Robinson is a very persistent fruitcake. He wrote the president, the vice president, all of us here, even a couple of the visiting speakers. I wish one of us had printed out the e-mail so that we could put it in the opening slide. But everyone deleted it, immediately."

Axel was about to say something, let them know that he, in fact, hadn't deleted the e-mail—but he could see from the looks in his fellow hobbyists' eyes that they were convinced that it was a hoax, and not worth their time.

Axel decided to keep his suspicions to himself. Most likely, he was the one who was overreacting, and his friends were correct—it was some nutcase, a fruitcake, a waste of everyone's time. But in Axel's mind, no matter how you looked at it, this was wrong. If the person behind the e-mail actually had moon rocks, he had to have stolen them. If he didn't, then he was trying to commit a fraud.

"A shame," Axel finally joked back. "We could have used the offer as the front page of our next newsletter. Maybe entice a few new members, one of whom might bring over a collection that doesn't look like something I could use to pave my driveway."

They all had a good laugh as Schnermeyer moved toward the slide projector, preparing to get the meeting started. The other members of the Antwerp Mineral Club had already forgotten about the e-mail, and the nutcase who called himself Orb Robinson.

But Axel Emmermann still had images of moon rocks dancing in his head.

. . .

When Axel finally returned home from the youth center, the house was already dark. He let himself in as quietly as he could so as not to disturb his wife and kids. A second late night in a single week was incredibly unusual for him, but he had a feeling this was just the beginning of unusual things. He briefly considered waking Christel to tell her exactly what he was going to do—but he didn't want to reopen that can of worms. Besides, he really didn't think what he was about to do could be dangerous. Although he couldn't be sure, he guessed there was an entire ocean between him and Orb Robinson.

He crept through the house as carefully as he could and made his way to the darkened living room. He didn't even sit at the desk; he just stood in front of the computer in the corner of the room and began to type. When it was done, he stood back from the computer. Bathed in the warm, pixelated glow from the desktop monitor, he felt his cheeks flushing red.

Everything in its place, everything in its way.

His hand was trembling as he reached forward and hit the send key.

18

I might be interested if the price is right . . .
Have you any proof that the goods are what you say they are?

Thad hunched over his laptop as he sat on the edge of his and Sonya's bed, trying to convince himself that he should just hit the delete key, send the little packet of electronic information into the black hole of nonexistence, forget that he'd ever gotten the response, forget about the whole mental game he was playing, forget about the lunar vault and that little door that led to the safe filled with lunar "trash." Because now it was beginning to feel less like a game and more like something real. Here, in front of him, was a response—from some guy in Belgium, a mineral collector, a rock hound, with the Hollywood-ready name of Axel Emmermann. Axel Emmermann seemed ready and eager to commit what he had to know was a crime—purchasing an illegal "multi-carat" moon rock from a stranger on the Internet. Thad was still playing a game, but this Axel Emmermann wasn't; he was really looking to buy a piece of the moon.

Thad ran a hand through his flop of curly auburn hair. He was wearing only a bathing suit, having just come back from a day of scuba instruction at the local Y. His hair was still damp, and he could feel the goose bumps rising across his naked chest and back. He was in no rush to get dressed, even though he was supposed to be getting ready for dinner. He really had no interest in going out with Sonya and her friends tonight—even before he'd gotten the e-mail from the Belgian gem collector, he had contemplated telling Sonya that he wasn't feeling well.

There was a peal of sudden laughter from the direction of the living room, and Thad glanced up at the closed bedroom door. He didn't know how many of Sonya's model buddies were gathered out there—when he'd come home from the pool, he'd counted at least four buffed and polished specimens, as well as at least three already opened bottles of red wine—but he didn't think he could handle another evening of mindless conversation in some loud, overpriced, overly trendy restaurant. And now, looking at this e-mail, he knew there was no way he would even be able to fake his way through the ordeal.

I might be interested if the price is right. . . .

Thad shivered. He hadn't really thought about price yet, even in the context of a mental game, because in truth, he hadn't any idea how one would actually manage to pull off the heist. The lunar vault was unbelievably secure, from the keypad that got you over from Building 31 to the monitored entrance that led on through. Then there was the clean room and, of course, the huge steel vault door itself. The "trash" safe held about seventeen pounds of lunar rocks, but Thad didn't think it would be possible to smuggle that much rock back through the clean room or past the security cameras. Which is why, in the original e-mail he had sent out to the dozens of foreign collectors, he had specified only a single "multicarat moon rock."

Working from that thought, he tried to figure out what kind of money he could ask for—how much imaginary cash he'd be demanding in return for his imaginary moon rock. Even though he remembered a single gram of lunar material was once put on the market for $5 million—and even if that number seemed ridiculous, he had read somewhere else that at a Sotheby's auction, a gram of lunar material once went for $400,000—Thad didn't intend to be anywhere near that greedy. He wanted this to be quick and easy, the kind of transaction that wouldn't draw any attention. A Belgian gem collector couldn't possibly have that kind of money, anyway; Thad needed to come up with a number that was both achievable and high enough to make it worth his while.

Make it worth your while, he repeated to himself, incredulous at his own thoughts. Breaking into a NASA building, stealing the most valuable thing on Earth, endangering his chances of ever becoming an astronaut—Thad shook his head. That was only one way of looking at the mental game. NASA had designated those rocks as trash, unusable. Thad could use the money to make himself a better scientist, a better candidate for the astronaut training program. He'd be out of debt; he'd have money to pay for schooling, research, whatever he needed. And if he became an astronaut, he might one day help NASA in its quest to get to Mars—which meant, in a way, this theft would be a good thing for the institution. He had to continue to think about the heist in those terms—because in those terms it was more than palatable, it was heroic and noble. Thad thought of himself as a scientist, and he would use whatever he earned from the heist to advance science. To advance himself, within the realm of science.

And besides, you couldn't get arrested for pulling off an imaginary heist, could you?

Thad dried his damp hands on the blanket beneath him and looked at his e-mail account. He dashed off a quick message to Gordon, asking him to investigate this Axel Emmermann, to compile whatever he could find via the Internet and whatever other means he had at his disposal. Thad wanted to know whom he was dealing with before he took the next step.

Meanwhile, he began to compose the response he would eventually send to the Belgian rock hound. He didn't even notice when Sonya and her friends exited the house, trailing laughter, mindless banter, the cacophony of clinking wineglasses. He didn't even notice as his green Toyota Tercel started up in the driveway, the tires spitting gravel as the group headed to the restaurant without him. Sonya hadn't even remembered that he was supposed to be going to dinner with them—but Thad didn't notice, and truthfully, if he had, he wouldn't have cared.

19

Up in the air. It's a bird. It's a plane. It's Emmermann . . .

Axel grinned as he pictured himself flying circles around the sparkling blue sky over Antwerp, his charming potbelly struggling mightily to break free of a bright red spandex costume, a silken cape billowing behind him as the warm spring air whizzed about his aerodynamically bald forehead. He saw himself winding low above the sixteenth-century churches and castles, the tourists waving and applauding as he showered them with moon dust . . .

His grin became an outright guffaw. *Axel Emmermann, superhero.* At the moment, Axel the superhero was on his knees in the little patch of dirt beneath the window of his living room, his face bright red as he fought a particularly villainous species of weed. He was using both hands in a patented throttle motion, pulling with all his strength as he attempted to dislodge the shock of bright green—a botanical brute that was strangling his wife's parsley at the root. The damn thing was hanging on for dear life, and it felt as if the nasty plant's tendrils were gripped around the very core of the Earth.

He wished Christel could see him now, on his hands and knees in the mud, at the mercy of a brainless twist of plant. She would have seen for herself how amusing the label she had given him was. He was as far from a superhero as a forty-nine—soon to be fifty—year-old rock collector could be. But Christel was gone; she'd stormed off to the market right after Axel had sent his latest e-mail, leaving him alone in the house to face the nefarious weeds.

Christel wasn't really angry; it was more a mix of frustration, and maybe a little fear. Because many days had passed since Axel had sent his first response to Orb Robinson, Christel had assumed that the matter had been dropped. And rightly so. The other members of the mineral club had long forgotten the foolish hoax. But Axel was a different breed, and even after a week he couldn't seem to let the matter go. Maybe he was more like a weed than a superhero.

That very morning, he had decided to take action. No spandex was involved. Just himself, in his gardening shorts, knee-high boots, and short-sleeve shirt, alone at his computer, just hours ago.

It hadn't been very hard to Google his way to the Web site for the Tampa, Florida, division of the FBI. He didn't know for sure if Orb Robinson really resided in Tampa, but he couldn't think of any better place to start. Certainly, the Belgian police force would not have gotten involved in a case of fraud like this. And he doubted that Interpol even had a Web site. Moon rocks were a uniquely American treasure, so if anyone should be investigating this, it was probably the FBI.

Axel had crafted the e-mail carefully. His only mistake was to show it to his wife before he sent it:

> I am a mineral collector who lives in Belgium. Some weeks ago I was contacted via e-mail by a person, Orb Robinson, who claims to have some lunar rocks from NASA for sale. He also advertises on "the Virtual Quarry" of our Web site with the following ad: "Priceless Moon Rocks Now Available!!! . . ."
>
> I seriously think that this person is trying to swindle unsuspecting people out of a lot of money. I have answered his e-mails indicating that I would be interested in a buy if the price was right. If you want, I

can forward these e-mails to you. I realize that this probably is a low-priority event, but nevertheless I would like to report it.

Best Regards, Axel Emmermann

Thinking back to Christel's highly vocal response to the e-mail when he'd told her about it, Axel finally got a good grip on the weed and heaved his not so insignificant bulk backward; the weed finally came loose, nearly sending him tumbling back into the mud. He caught himself at the last minute and tossed the weed into a garbage pail. Then he rose heavily, his aging knees creaking with the motion. He shook the dirt from his bare legs and pulled off his muddied gloves. Then he headed back into the house.

As he approached the computer in the corner of the living room, he wondered if Christel was right. Not about him being a superhero, or about the danger of getting involved, but her observation that this really wasn't about right and wrong, that Axel couldn't let this rest because this, to him, was fun. It was a game, another hobby. Like popinjay, except instead of shooting arrows at a wooden bird affixed to a pole, he was casting e-mails at an invisible foe.

It was true, he wasn't working at the moment, having taken a short disability leave from the polyethylene plant where he was a quality control supervisor, due to a recurring injury, and rock collecting could only take up so many hours during the day. Archery filled most weekends—but maybe fighting crime would make up the difference.

No doubt he felt a surge of adrenaline as he sat down in front of the computer. He wasn't really expecting the FBI to answer him so quickly, but he had a feeling he'd be checking the computer more than a few times each day. Sooner or later, he would get a response.

To his surprise, the minute he opened his account, he saw that there was a new message in his in-box—but not from the Tampa division of the FBI. Quite coincidentally, the e-mail was from Orb Robinson:

> Yes, valid proof will be provided. What is the approximate range of $/gram that you consider "right"? Let's discuss your possible interest and see if we make a great business partnership. If you are truly interested then, I will provide you with more detailed information.
>
> Sincerely, Orb Robinson

Axel couldn't believe what he was reading. After more than a week, and directly after he had contacted the authorities—now this Robinson had finally answered him. Like the weed in his garden, the nutcase hadn't just given up and gone away.

Looking more carefully at the new e-mail, Axel immediately noticed something. Robinson was no longer talking about carats; now he was talking about dollars per gram. To a rock collector like Emmermann, it was a significant difference. Exactly how many moon rocks did this character have? Did he really think that a mineral expert would believe he had in his possession many grams of the rarest substance on earth?

Axel knew one thing for sure: he wasn't going to wait for Christel to get back from the market to craft his response.

He quickly came up with some numbers, almost off the top of his head. He was playing the role of an interested buyer, so he had to keep it believable.

> Hi Orb,
>
> If you can provide valid proof that these rocks are really lunar samples, I would be willing to buy if the price does not exceed 800$/gram for rocks under 10 grams and 600$/gram for larger specimens.
>
> Axel Emmermann

. . .

It wasn't until the next day that Robinson responded. At around nine in the morning, Axel was shuttling back and forth between the kitchen, where Christel was serving his son and daughter oversized waffles painted in more butter than was nutritionally safe, and the computer, checking his e-mail again and again—and then, there it was. Axel no longer even attempted to hide the burst of excitement that exploded across his face as he saw the new message:

> Axel Emmermann,
>
> Your prices are just fine; in fact I can do better than that, but I have minimum mass requirements. To give you an idea of my mass range, I would prefer to stay around 1 kilogram. The following is the breakdown of the varying price range . . .
>
> $500/g (.5–.64 kg)
>
> $400/g (.65–.85 kg)
>
> $300/g (.86–1.5 kg)
>
> Of course, verification will be provided before you purchase, I think that if you are seriously interested then we should meet and confirm this deal in person. Please let me know what you think.
>
> Sincerely, Orb Robinson

Axel leaned back in his chair. Carats had become grams, and now grams had become kilos. Christ, this hoaxer was brazen. Axel quickly did the calculations in his head. At the prices Orb Robinson was quoting, one kilogram of moon rock would cost around $300,000. It was nowhere near what real moon rocks were worth—but it was an enormous amount of money to a man like Axel. Even talking about trading that kind of money for an illegal specimen caused his crime-fighting hackles to rise. This wasn't a little hoax—it was significant.

Reading the e-mail again, Axel began to think that maybe he had taken this as far as he could go. He now had this hoaxer quoting prices, and the only thing left was for him to fill a suitcase, hop on a plane, and head to Tampa. Of course, he was never going to do that. In forty-nine years, he had never been far outside of Belgium. Certainly not to the United States. And he had no interest in buying what was most likely a huge chunk of fake moon rock. If this was going to go any further, someone else was going to have to take charge.

. . .

Two days later, Emmermann was back at the computer when the cavalry finally came riding in. He clicked open the e-mail as soon as he saw the header—from the Tampa division of the FBI.

> Mr. Emmermann:
>
> First, thank you for forwarding this information to me. You have piqued my interest.
>
> Second, I'm afraid I am somewhat less than familiar with the laws surrounding the sale and/or possession of moon rocks. I assume—based on the letter from Robinson—that it is probably illegal. In fact, I would guess that Mr. Robinson is either in possession of contraband or is misrepresenting (in an effort to defraud someone) a more mundane mineral, i.e. he is violating the law in one way or another. Could you please let me know if my assumptions are correct?
>
> Last, if we do initiate an investigation into this matter, would you be willing to introduce an investigator to Robinson as your representative in the States? Since Robinson has contacted you already, your credibility with him must be sufficient to quell any concerns he may have in conducting an illegal transaction with a complete stranger.
>
> Again, thank you for your alerting the FBI to this matter.
>
> SA Lawrence A. Wolfenden, Tampa Div'n

Axel was initially surprised that the agent from the FBI seemed to be relying on his interpretation of the situation—that he was basically asking Axel for advice on whether this was something that was worth involving the FBI. At the same time, Axel felt a huge gush of pride. The FBI was contacting him from all the way across the world. His wife might make jokes, but he really was doing something real, putting something back in its proper place. If she wanted to call him a superhero, well, now he had something to show her. But before he printed out the e-mail to run around the house with, he crafted his response.

> Mr. Wolfenden,
>
> I believe that your assumptions are indeed correct. The chance that Mr. Robinson has lawfully acquired real samples of lunar rock from NASA is, in my opinion, next to impossible. Therefore he must be in violation of at least a few laws. The tone of his message also suggests to me that this is not his first attempt to swindle some gullible overseas buyer. I may be ahead of things but I wanted to see where this would lead to.
>
> I would be more than willing to introduce an investigator to him if you are willing to investigate this further. In my opinion it would be better that you draft a reply since you're much more experienced in dealing with these kind of people. A real mineral collector (I'm not sure you have one of those on your team ☺) would express at least some concern about the verification. Of course, if I really was so gullible as to believe Mr. Robinson, I would be easily persuaded to buy if "my brother-in-law" was allowed to "take a look" at the rocks before buying. Since he already dove under my suggested price per gram, but his "multicarat" rocks have evolved to boulders of a half kilo and more. This would be my reply if I was really an interested buyer . . .

At this point in the e-mail, Axel drafted what he might say to Robinson if he was really going to go through with the transaction—but in a way to allow someone from the FBI to take over the situation, pretending to be his brother-in-law. Axel was really enjoying the creative aspect of this; it was as if he were a member of the FBI, plotting to bring down a master criminal. Of course, he didn't really think Orb Robinson was a master criminal, just a nutcase trying to pull off a hoax. Nonetheless, it was gratifying work.

Again to his surprise, the first response he got, later that day, wasn't from the FBI. It seemed that Orb Robinson was getting impatient.

> Axel,
>
> Please indicate if you are interested and/or able to purchase a rare lunar piece. Timing for me is sensitive, so I won't waste any of yours. We both know that you would be getting a great deal, and if you are still worried that I am trying to sell you a fake, good, I don't want you to just take my word for it. Please let me address your voiced concerns. Just let me know what they are. Acquiring this specimen is a sensitive matter for me, as you can imagine, and that is why I have the minimum mass requirement. It is more a minimum financial barrier that makes this transaction worthwhile for me and my group. So if you are skeptical about the validity of the origins of the rock, good, I shall provide you with convincing evidence, when I believe that you are serious. If you are concerned that you cannot afford this transaction, I understand. Perhaps you could find a significant number of customers that would be interested in purchasing pieces of your lunar sample and then would have the incentive to make such an investment. Either way, even if you are no longer interested, please indicate it to me.
>
> Thank you, Orb Robinson

Now this was truly fascinating. Orb Robinson had now become "me and my group." And even more interesting, Robinson had implied that the rocks had yet to be acquired. He didn't have the moon rocks in his possession? It was something he was going to get, somehow? The size of the specimen had remained in the kilo area, but this seemed very significant. If this wasn't just a hoax, if this Orb Robinson really was going to acquire moon rocks—then this was a crime that hadn't yet happened.

Axel ran a hand over his bald head. He had seen plenty of Hollywood movies, and he had always enjoyed the cat-and-mouse games played out between the cops and the robbers. Still, he did wonder—if he was offering money for these rocks, and that made Robinson go out and do something crazy to get them—was he actually inspiring the crime?

Axel shrugged his meaty shoulders. Robinson had already done something illegal. He had endeavored to sell moon rocks over the Internet. Whether they were real or, more likely, part of a hoax—it wasn't right. And if Axel hadn't responded, maybe someone else would have. Axel had done everything right. He had contacted the authorities. This Orb Robinson seemed very eager to make the deal happen. He was the one pushing it along, he was the one sending out the e-mails. He was the one committing the crime.

Axel wondered if Christel would agree. Soon he discovered the FBI certainly did.

In a rather long e-mail, Special Agent Wolfenden told Axel exactly what sort of response he wanted Axel to send back to Orb Robinson. It was a play on the draft that Axel himself had written, which filled Axel with more pride. His creative crime-fighting juices had obviously been accurate. Any ethical questions disappeared as he read through the e-mail. His "brother-in-law" had been changed to a "sister-in-law," obviously because the FBI had a woman agent they wanted to use in the setup. But all in all, it was as Axel had planned:

> Hi Orb,
>
> Your prices are better than I hoped for but the specimens are quite large. You spoke of "multicarat rocks" but a 500 gram rock would cost me 250,000$ and that is no small change. This amount is far out of my league and I would have to find one or more financial partners. Are these really the smallest rocks you have? I would be more interested in smaller specimens. I could always split up a larger rock for resale but it still is a large investment.
>
> Nevertheless, I can free 100,000$ on reasonably short notice. I would be happy to spend it on a single (authenticated) rock of at least 250 grams.

I can't free myself from work right now, so a meeting in person would have to wait until September. However, my brother and his wife live in Pennsylvania, U.S. I trust them completely and my sister-in-law is somewhat of a hobbyist in mineral collecting. She might be able to verify the rocks' origin, I think. Would you be willing to deal with me through her?

Sincerely, Axel Emmermann

It wasn't even a full day later when Robinson responded. Axel immediately forwarded the response to Agent Wolfenden. He felt a little like he was watching a movie in real time, played out over the Internet, these e-mails bouncing back and forth from the United States to Belgium and back to the United States. Who knows, maybe the criminal and the FBI were only a few miles apart, both communicating via a rock hound in Antwerp? It was the most exciting time in Axel's life since his days in the army, even though he was doing little more than sitting in front of a computer screen in his living room.

Axel,

In attempt to keep things out in the open between us, I will address my concerns. As you well know, it is illegal to sell Apollo lunar rocks in the United States. This obviously has not discouraged me since I live in the United States. However, I must be cautious that this deal is handled with delicacy in that I am not publicly exposed. This same law that makes it illegal to sell Apollo lunar rocks also for our mutual benefit makes them quite rare and valuable. My projected return from this has been just over $250,000 and I would of course prefer to be involved in one business dealing and get it all over at once in order to minimize my personal exposure. Having said that, if I can build some more trust with you, then perhaps I could do a deal with you for $150,000 and then if you find enough buyers you could buy the rest from me. As you can see this decreases my safety and increases my exposure and therefore I would only feel comfortable in doing this if I learned to trust you, which is difficult to do under the circumstances. Maybe you should give me the names of your relatives/contacts in the United States and then have them e-mail me and we shall begin to build a level of trust from that. I could meet them in the United States and then settle our mutual concerns and verify the authenticity of the specimens through them. I can acquire three very unique and valuable specimens and I am waiting to provide you with the details about them until I have built some more trust. One of them does involve dust. Please let me know if it is impossible for you to find some more investors in order to make this in one purchase. I would prefer that over two purchases. Either way I am interested in developing this business relationship with you. And I wish you a hefty profit from our encounter. Please reply with your thoughts and or concerns.

Thank you.

Sincerely, Orb Robinson

Axel contemplated Orb Robinson's tone as much as the message itself. It seemed the little hoaxer was getting frustrated with Axel's limited funds—and he also seemed very eager to get this thing done. He also mentioned "three very unique and valuable specimens," which seemed specific. If this was a hoax, why would he make any specifications at all?

Axel fought the urge to respond on his own, waiting until the FBI sent him a draft of what he was supposed to say:

Mr. Emmermann:

Following is the reply we'd like to send to Robinson:

Hi Orb,

I would prefer to make the first purchase at 100,000$ as we have discussed. If the lunar rocks are proven authentic and all goes well, I will have a much easier time of convincing others to invest and help with a second purchase. I have spoken to my brother and sister-in-law and they would be willing to purchase the lunar rocks for me. As I may have told you earlier, my sister-in-law is a hobbyist in mineral collecting. She has allowed me to provide you her e-mail address, which is xxxxxxxxxxxx@xxxxx.xxx and said she'd be willing to stand in for me in this initial transaction. Although I trust my sister-in-law, I do not necessarily trust her abilities completely. How will you provide that the lunar rocks you offer are real? Can you provide me some documentation as well? Are they meteorites or samples from one of the Apollo missions? I would not be interested in purchasing meteorites.

Sincerely,

If Robinson likes this, I think the next message will probably be directly to myself and Agent Nance. I'll let you know.

SA Lawrence A. Wolfenden, Tampa Div'n

Axel understood, reading the new letter as he forwarded it along to Orb Robinson under his own e-mail identification, that from here on out, if things went well, Robinson would be contacting his "sister-in-law" directly, and the FBI would probably be able to take over from there. There was a sense of deflation as he realized that he was giving up his front-row view of the investigation in progress, but really, there wasn't much more he could do from Antwerp, and he wasn't about to jump on a plane to meet face-to-face with a master criminal. From the sound of the e-mail, it seemed the FBI was going to put together $100,000 to try to entice this Orb Robinson to make the deal. Agent Wolfenden seemed to be taking this quite seriously.

The next e-mail from Robinson, just one hour and eighteen minutes later—the last one Axel would receive for quite some time—made it clear that no matter how seriously Agent Wolfenden was taking the situation, it wasn't any sort of overkill. If this was a hoax, Robinson was going to take it right up to the edge of the precipice, right up to the exchange of money for rubble:

> Axel,
>
> Since I am confident of the authenticity of these rocks, I will hope that you are able to find many customers quickly after our first transaction, and will for now continue planning on making this transaction. I will e-mail your sister-in-law and begin setting up a meeting time and location. You make sure she is prepared to pay in cash, and I'll make sure that she has all the relevant documents and publications on the individual specimens. The type of proof I will be providing will be the scientific publications, which can be easily verified and reproduced by you. In these documents/publications, there are quantitative measures describing the samples, photos, and unfakable descriptions of them. I encourage you to have your sister-in-law bring all the scientific equipment she has access to if she wants to double and triple check the samples for the accurate properties. I cannot alert you as to which exact samples will be involved before the trade, because the exposure becomes too high. However, I understand that during a transaction she, and I assume that her husband will be there for her protection, will want to have ample time to check the samples before purchase. So we will discuss the details of the transaction at great detail before it takes place. Please continue to stay in touch with me, and inform me of any changes, concerns, or updates.
>
> Thank you.
>
> Orb

Reading this last e-mail, Axel had to admit that this was no longer sounding like an elaborate hoax. Maybe his wife had been right to be concerned; maybe there really was some level of danger in this can of worms. Axel was glad that he had brought the FBI into this—that the real authorities were going to handle it from here. Because superhero or not, Axel didn't feel like he was playing a game anymore. This was beginning to feel like something that was deadly serious—and whatever was going to happen, it was going to happen soon.

Axel was now pretty sure of one thing.

Orb Robinson was about to commit a major crime.

The seconds slowly tick away. They roll into their infinite repetitions echoing the never-ending groundhog day. The cold concrete is eternal.

20

JSC, June 6, 2002

Thad pressed his rubber heels against the curved, brilliant white surface of the International Space Station and pushed off, feeling the sudden rush of adrenaline as his body floated forward. His arms instinctively rose out from his sides, his gloved palms outstretched, and for a moment he was like some sort of wingless angel gliding through a devastatingly blue void. Then his body started to spin, somersaulting forward on its axis, a slow-motion human pinwheel riding out a carefully modulated arc, moving farther and farther away from the vast hull of the station in a symphony of weightless motion. Almost immediately, another shape began to grow at the edges of Thad's revolving vision. Expertly using his arms to slow the revolutions, in a moment he was once again nearly still, floating upside down in the empty blue. He stared out through his Plexiglas faceplate at the shape that was now a form.

Even from his upside-down vantage point, the space shuttle was a thing of beauty. From his angle, Thad could make out only a portion of the cylindrical cargo bay, but his mind easily filled in the rest. From the curved, sleek nose cone slicing through the blue somewhere up ahead to the jutting shark fin of a tail towering upward—just out of view—the shuttle's muscular presence was entirely palpable, even as it hung frozen nearly fifty feet away. Thad was so transfixed by the sight of the thing that he didn't notice the hatch embedded halfway down the fuselage lifting open until it was nearly perpendicular to the ship itself.

He couldn't make out any details at that distance; the interior of the cargo hold was nothing but a dark yawn. But suddenly a new shape appeared in the darkness, rising up to fill the open hatch like some sort of alien creature. Bulbous, an even brighter shade of white than the International Space Station behind Thad—the creature had legs and arms but was obviously more machine than human. Its legs were thick like tree trunks, ending in enormous, rubber-soled boots. Its arms, almost as thick as the legs, were stretched outward, gloved hands gripping the sides of the open hatch as if readying for a huge leap forward. The white torso of the thing was a pincushion of tubes and hoses, running around both sides to a giant rectangular pack attached to its back.

Still suspended upside down more than fifty feet away, Thad lifted his gaze to the machine creature's face. Except . . . it had no face—instead, where its face was supposed to be, he found himself looking at a curved sheet of reflective, gold-hued material, polished so spectacularly smooth that it glowed as if lit by its own internal light source.

"My God," Thad whispered.

It was not the first time he had laid his eyes on an EMU, but it was the first time he had seen one like *this*: fully operational, worn by a real astronaut as he was about to step out of the cargo bay of the space shuttle. The Extravehicular Mobility Unit was a particularly fancy name for a space suit—but to be fair, it was a *very* fancy space suit. More like a spaceship, actually, a completely self-contained unit designed to protect the astronaut from the harshness of space. Its construction derived from the original design that had been used during the Apollo missions, the EMU was one of the most sophisticated tools in the NASA arsenal. From its hard upper torso made of fiberglass, containing the control module and the primary life-support systems—bleeding the tubes and hoses that controlled oxygen, and cooling and warming liquids that kept the astronaut alive—to the ultrasophisticated helmet, composed of a vent pad that controlled the flow and pressure of oxygen, to the recognizable bubble, which was covered by the extravehicular visor assembly, coated in a thin layer of pure gold to filter out the sun's dangerous rays.

Aside from television and movies, very few people ever got to see an EMU in action, and here Thad had a front-row seat. It was a spectacular, frozen moment in time, and he wished he could have hung there forever, upside down in the blue vastness, lost in his own mind, his fantasies. Because in his fantasies it was him in that EMU, staring out through that gold-tinged visor, stepping forward through the shuttle cargo door into the nothingness of space. It was him in the astronaut suit, beginning a space walk a few hundred thousand miles above the Earth, working his way toward the International Space Station, joining the ranks of the heroic men and women who had ventured into space with the whole world watching. It was him—not a poor Mormon reject turned out of his home by his own parents, in a rapidly failing relationship, a pretender in a personality and a place he didn't really belong—a real live astronaut about to do something dangerous and meaningful, to make his mark on the world. And then, seemingly out of nowhere, a voice reverberated through him, bringing him back to himself, hanging upside down in the infinite blue.

"Shuttle Commander, this is mission control. Proceed with EVA. Delta Alpha, on our command."

Thad shivered as the words moved through his skull. It was an odd feeling, sound translated directly through bone, as if the words were coming from inside his body instead of from the cheek transmitter that extended out from his face mask. The bone conductor was another cool NASA toy, and even though Thad could only receive sound and not respond, it added a whole new level of sci-fi to the moment he was viewing.

"Affirmative, MC. On your mark."

The response was slightly accented, maybe Texas or Louisiana, but still Thad couldn't quite place the voice. He didn't know the astronaut stepping out of the shuttle bay door personally, but if he got close enough he was sure he could identify the man through his helmet. That thought alone was thrilling enough; even though he was one week into his third and final tour, Thad hadn't yet met enough astronauts to make them any less godlike to him. And now, watching one in an EMU, in full action, it was no wonder, because they truly did seem like gods. He wished he could say something, that the bone conductor transmitter worked both ways, that he wasn't just a voiceless observer. But then again, what would he say? Despite his fantasies, he was just a co-op.

"And go."

There was only a brief pause, and then the space-suited figure launched himself forward off the bay door. His bulbous suit floated forward at about twice the speed Thad had originally launched himself off the space station. It was spectacular to watch. And it wasn't a dream, it wasn't a fantasy—but then again, it wasn't exactly *real* either.

"Roberts"—a different voice suddenly broke through Thad's bone conductor—"nice job attaching the strobe camera to the station hull. Go ahead and break surface. Doc will check you out, and you can hit the showers."

Thad sighed, took one last look at the astronaut in the EMU—then spun himself around and kicked off with his rubber flippers. He wriggled upward, feeling the cool blue water rushing against his rubber wet suit. Even though he was breathing nitrox and not regular compressed air, he still had to be careful not to rise too fast. The nitrox—a mix regulated to control the nitrogen uptake into his body to extend his dive time—would offer some protection, but there was always the risk of the bends. He had been down for a long time, getting the camera affixed to the station hull just right so Mission Control could photograph the simulated space walk; it had been painstaking work, made all the more difficult by the near-zero-gravity atmosphere at that depth, and even with the flashlight attached to his breathing mask, it had been hard to see because of the strangely, and overwhelmingly, blue nature of the deep-water surroundings.

Thad was pretty well exhausted, but still he moved with the precise control of an expert diver, rising foot by foot with the slightest effort from his legs and fins; finally, his head burst out through the surface of the water, the blue vanishing from his eyes in a blast of bright fluorescent lighting. It took him another few minutes to swim the twenty yards to where Brian Helms was standing, looking down with a grin of approval on his face. Brian's own wet suit was unzipped, hanging down around his waist, revealing a bare chest that was almost as many bony angles as his triangular face. Like Thad, still in the water, Brian was breathing hard. He had only just climbed out of the pool himself, moments ago.

To Thad, calling it a "pool" was more than an understatement: the Neutral Buoyancy Lab—the NBL—was simply massive, and one of the most impressive facilities at NASA—though it wasn't located on the JSC campus, instead being housed in the ultrasecure Sonny Carter building, a ten-minute drive through Clear Lake from the main campus. The NBL was the largest indoor pool of water on Earth. Two hundred and two feet in length, 102 feet wide, 40.5 feet deep, it contained over six million gallons of water. It was the premier astronaut training environment in existence.

As Thad looked back across the vast expanse of water, it was almost hard to believe that deep below, where he had just been, there was a full mock-up of the International Space Station, the space shuttle cargo bay, and even the Hubble Telescope. The only clues to what lay below were the enormous, bright yellow mechanical cranes that hung down into the water, which were used to reposition the mock-ups for different projects and training programs. Even though Thad had been working in the NBL for six days now, he was still in awe of the facility. When he had first come to NASA, he had read about the NBL, but seeing it in person was a truly humbling experience. Because in many ways, like the Space Shuttle Simulator he had snuck into during his first week, this was as close to space as a nonastronaut could get.

Deep in the pool, astronauts could practice space walks—as Thad

had just witnessed—as well as work within the space station and shuttle in a neutrally buoyant atmosphere. The EMUs protected the astronauts from the water as well as the pressure, and with the help of divers to keep them properly buoyant, they got a pretty good approximation of the real sensations of living and working in zero g. Underneath the EMUs, the astronauts wore long johns and a body diaper, as well as the liquid cooling and heating tubing that kept them at the proper temperatures. Gloves, regulators and fans, helmet coms—the EMU contained everything the astronaut would be using in a real space walk, kept at a perfect internal pressure of 4.3 psi. Dressed like that, an astronaut could live in the suit for almost nine hours, and though the experience was slightly altered for the underwater environment, the concept was strikingly similar. Any mission that took place in space began here, and now Thad was part of all that, working on projects that would one day be replicated in outer space.

Thad found himself grinning just like Brian as he pulled himself out of the pool and onto the wide, grated metal deck of the NBL. His body felt strained, and he could hear his heart pounding in his chest. He had been down a little longer than he had realized. Because of the nitrox, it was unlikely he would get the bends, but even so, after the dive doctor checked him out and sent him to the showers, he'd have to wait around for a couple of extra hours and be rechecked to make sure the time under water hadn't done any damage. But truly, he didn't care about the minute physical risks of being a dive assistant at the NBL; he wouldn't have traded the position for anything. It was the plum co-op job, the most-sought-after posting in the co-op program. The fact that he was there, on the NBL deck—and that Brian was with him—was pure luck. Or really, more timing than luck—like a lot of things in Thad's life, it was about being in the right place—and knowing how to take advantage of an opportunity.

• • •

After two tours spent in the life sciences department, Thad had been looking to do something different, something more active, something where he could take advantage of his unique skill sets to further impress the NASA brass. So when he'd heard about the NBL position opening up, he had been instantly intrigued. That Brian—who had decided to return to NASA for an unusual fourth tour at the behest of his mother—was interested in the position only added to the spirit of the competition. Usually, there was a two-year waiting list for NBL jobs, and no doubt there would be dozens of applicants aiming for the same spot.

The week before Thad was supposed to go for his interview, he was stepping out of his usual Tuesday-night volunteer firefighter meeting—Wednesdays were sailing; Thursdays, rock climbing—when he noticed something on the sidewalk in front of him. A wallet, obviously dropped by someone by mistake. Inside the wallet was an ID with a picture that Thad vaguely recognized from previous firefighter nights—but not someone he knew by name. There was also a NASA ID—so as soon as Thad returned home, he looked the guy up in the NASA directory and made the call. Getting an answering machine, he had left a simple message.

An hour later, the man called back, thrilled that Thad had found his wallet. He had already canceled his credit cards and was working on getting a new NASA ID, but Thad still volunteered to drop the wallet by wherever the man worked. Instead, the man scheduled a time to swing by Thad's lab in Building 31 the next day.

When he pulled up in a little red convertible sports car at the prescribed time, Thad was surprised at how appreciative and thankful the man appeared to be; Thad hadn't done anything anyone else wouldn't have. But in any event, by the end of that day Thad had forgotten about the exchange. In fact, he didn't even remember the man's name.

A week later, he and Brian did their best to remain optimistic as they carpooled the short ten-minute drive over to the Sonny Carter Training Facility for their NBL interviews. They were both excited just to see the building, as it was one of the highlights of the institution, and neither one of them had clearance to get inside without an escort. They didn't think they'd actually get a chance to walk on the deck of the great pool itself, but just being in the same building where the astronauts trained underwater would be an amazing experience.

When they stepped inside the building, they were directed to a small waiting area, just outside the heavy doors that led into the main part of the facility. Not surprisingly, Brian was the first one ushered down the hall toward the administration offices. There was no doubt, now, that his mother had made a phone call at some point down the line. Thad didn't begrudge his friend his connections; if Thad had been lucky enough to have been born to a NASA engineer, he would have carried the man or woman's picture in his pocket everywhere he went.

Twenty minutes later, Brian made his way back into the waiting area—a huge smile on his face.

"Turns out we went to the same college," he said, barely containing his joy. "So we had a lot to talk about. I think I'm in."

Thad gave him a high five; he was genuinely happy for his friend, not just because he liked and respected Brian, but because if Brian got clearance to the NBL, maybe he'd be able to bring Thad inside along with him, every now and then.

It was another twenty minutes before Thad got called for his own interview. Brian wished him luck, and then Thad was moving quickly down the hall that led to the administrator's offices. The designated room was at the end, the door already open. He stopped in the doorway, peering into what looked like a huge space with an oversized wooden desk in one corner. There was a man sitting at the desk, back to the door, typing away at a computer. Thad immediately got the impression that this man had already made his decision. He certainly didn't seem eager to meet another co-op.

But when the man finally turned around, Thad was in for a shock. The guy was grinning just like he'd been a week earlier, when Thad had returned the lost wallet to him. The man's name was Mike, and it was just one of those crazy coincidences in life; Mike was one of the NBL's project managers, and he had been given the task of hiring the new co-op for the dive assistant position.

"We can end the suspense right now," he said, "because you've got the job. You, and your buddy out there; I'll have to shuffle a few things, but we can certainly make room for two. But we've got to sit here for at least fifteen minutes and make it look like we're doing an interview so I don't get in any trouble."

And for the next fifteen minutes they chatted, mostly about the NBL and how Thad would have to get recertified in scuba to NASA's exacting standards. Thad wasn't worried about any testing; he had yet to fail at anything he'd gone after.

. . .

"I don't care how cool that monster of a swimming pool is," Brian shouted over the sound of the space-age shower jets that were pummeling him from every angle. "The thing that really blows my mind is the goddamn towels."

Thad grinned, raising his face to let one of his own shower nozzles go to work on his neck and upper chest. The tension was bleeding right out of him as the superheated jets of water cycled from soft to hard in a prearranged massaging program. The astronaut shower room was totally a scene right out of *The Jetsons*. The shower cubicles themselves were single-unit pods formed out of some seamless, space-age material. The electronic control panels affixed to the smooth interior walls were incredibly complex, offering control of the water temperature as well as the nozzle pressure. Thad was amazed at how extreme on either end the temperature and pressure could get; you could almost burn the flesh from your bones if you wanted. And since he had plenty of time to kill, waiting for the dive doctor to release him from the facility, Thad liked to set up massage routines that utilized a really wide range of pressures and temperatures. The controls also allowed him to choose the type of shampoo and conditioner he wanted to squirt out of the dispenser knobs attached to the ceiling. It was not unusual for a dive assistant's shower to last half an hour, or more.

And when Thad stepped out of the shower—that's when the mystery set in. Like clockwork, a steaming hot towel would appear out of the wall in front of him. Both Thad and his friend had spent hours searching for the sensors that told the computer it was time for the towel—to no avail.

"I think it's something they brought back from space," Thad responded as the streams of water pummeling his body finally softened, indicating the end of the cycle. "Definitely alien technology."

When the water completely stopped, he stepped out of the pod—and there it was, the mechanical whir followed by a hot towel. He grabbed the towel and wrapped it around his waist. Brian was already at his locker, retrieving his NASA shirt and khaki pants.

As Thad approached the locker next to his friend, a thought sprang into his head—and certainly not for the first time. He and Brian were pretty close, had worked next to each other on and off over a long period of time. Brian didn't really know him—just the persona he put on at NASA, the character he had become since that first evening at the pool party. But he did consider Brian a pretty good friend.

He wondered: What would happen if he told Brian about the e-mails, and the mental game he had been playing? But then he quickly shook the thought away. As much as he was dying to tell someone other than Gordon—who really was little more than an acquaintance—Brian would never have understood. Brian wouldn't see it as a mental game, or some sort of potential prank—which is how Thad was beginning to describe it to himself. Brian would see it as a potential crime.

Even though NASA considered the rocks trash, and there was no good reason for them to remain stored away, forever, in the darkness—well, maybe it *was* a potential crime. But if Thad somehow figured out how to pull it off—what a fucking crime it would be.

Because the truth was, even though Thad had made himself sound so confident in the e-mails he had exchanged with Axel Emmermann—and since then, the man's sister-in-law, a woman named Lynn Briley—the truth was, he had no idea how he would even begin to pull something like that off. The lunar vault still seemed pretty much impregnable to him.

He'd made some headway coming up with some of the steps he'd need to complete, some of the preparation he'd need to engage in before he even got started—but overall, he still didn't have a full idea of how he was going to get to that safe full of discarded moon rocks.

All the more reason why he wished he could talk this out with someone, to bounce it off a confidant. But watching Brian at the locker, pulling his NASA shirt over his triangular face—the guy was too straitlaced, too NASA to the core.

For the moment, despite all the e-mails, despite the fact that there was now a woman in the United States who had gathered together $100,000, ready to pay him—a massive sum of money to Thad, who had never even imagined having that much cash—this would remain a mental game, another fantasy like the dozens of other fantasies that made up the texture of his daily life. An impossible, wonderful, terrifying fantasy.

Then again, a few feet away there was a swimming pool that contained the International Space Station and the space shuttle. A few feet away, astronauts in EMU suits conducted space walks while Mission Control guided them along, with the help of bone-conducting transmitters.

At NASA, nothing was impossible.

And a hundred thousand dollars—wasn't that a pretty good motivation to solve the problems that lay ahead? Was the money worth the risk? Thad was on his third tour. He was working in the NBL. He was a star co-op. There was a chance he'd be hired by the JSC after he graduated.

At the same time, his relationship with Sonya was almost over. He wasn't even sure he was ever going back to Utah. Utah seemed like a million miles away. NASA was his entire life, even if, as a co-op, it was still more fantasy than reality.

One day, couldn't this be his life for real?

A hundred thousand dollars couldn't be enough to make him risk all of that, could it? There had to be something else—a catalyst. Something to transform this from a mental game into something else. Gordon had been a first step: a link who'd gotten him in touch with someone willing to put up the cash. But to make this real—he needed an even more significant trigger.

Without a truly powerful new catalyst, to clear away the fog of fantasy, this would never be anything more real than a space shuttle at the bottom of a swimming pool.

21

In retrospect, the flip-flops were a mistake. Thad had managed the hour-and-a-half drive south all right, his Toyota leading the small caravan of mostly expensive foreign cars from the JSC outer parking lot to the Texas coast without losing a single co-op along the way. But once they had all jammed their way onto the single-deck ferry for the short trip to their Galveston Bay isthmus destination—the cars packed so closely together there was barely room for the doors to open the necessary few inches to allow the more adventurous among the group to slither out during the short ride over to the campsite—Thad realized he ought to have chosen more appropriate footwear. EMU space boots with magnetic grips would have fit the bill—although he would have settled for his dusty Timberlands.

He stumbled forward as the ferry pitched hard over an errant wave, and barely caught himself before tumbling halfway over the hood of a black BMW sedan. There was maybe a foot and a half between the BMW on his right and a Range Rover to his left, and he was forced to stutter-step as he moved between the cars. The flip-flops were playing havoc with his usually perfect sense of balance. The soles of the damn things kept getting caught on the metal ridges that marred the floor of the ferry, and it was unlikely he was cutting much of an impressive figure as he made his rounds up and down the line of cars, greeting his co-op followers. It wasn't like he'd prepared a speech or anything, but he did have a reputation to uphold. A lot of the first- and second-tour kids didn't know him personally, but most had signed up for the weekend excursion because Thad Roberts, social star of NASA, had arranged it—which meant it would be something fun, something different. *An adventure.*

He reached the driver's-side window of the BMW and shook hands with the young man inside, introducing himself. He filed the guy's name away in his nearly photographic memory, then continued on toward the next car. Another wave pushed him back against the rear of the Range Rover—so he improvised, giving high fives to the young couple in the backseat of the vehicle, second years he recognized from the JSC cafeteria—and then he steadied himself using the bumper of a four-door Mercedes.

The Galveston Bay excursion was one of his most popular, which was the reason he had chosen it for the first weekend of his third tour. The beaches where they were headed had pretty relaxed rules—which meant that by nightfall, Thad would have multiple bonfires up and going, with no worries that any authorities would come by to make them put them out. There would be alcohol, of course, though Thad hadn't brought any himself and wouldn't partake; he'd probably never be able to look at booze or cigarettes or drugs of any sort without being reminded of his father's vivid stories of hell and damnation—but he enjoyed being around the party atmosphere that alcohol usually inspired. Not that alcohol or the bonfires would be the highlights of the excursion; in this case, nature was going to top anything he or his charges could arrange.

The beach they were going to was known for more than a lax police presence; where they were going, it was all about the algae. Thad had always believed that bioluminescent algae was something you had to experience for yourself; there was nothing like wading out to your waist, churning your hands along the bottom—and watching the water light up like the Fourth of July.

And as if the bonfires and the algae fireworks weren't enough, Thad had something even more spectacular planned for later in the weekend. He was going to cap things off with a truly dramatic adventure that the co-ops would remember for the rest of their tours.

He was having a little better luck with his footing as he passed the Mercedes and a second BMW, glad-handing the co-ops who filled each of the vehicles. He could tell from the progress of the ferry that they were only ten minutes away from their destination, and he was about to turn back toward his car when he noticed the Jeep Cherokee at the front of the row to his left, only two more cars down. He figured he might as well make it to the end, say hello to everyone; besides, he knew the driver of the Jeep, a second-year co-op named Chip Ellis who had been with him on a dozen other excursions in the past.

He was halfway to the Jeep when, looking through the oversized vehicle's back window, he noticed that there were two passengers in the rear seat. Girls he didn't recognize, one of them tall with light hair, and the other, in the farthest seat over, petite, with sable hair cropped short, almost a pageboy cut.

Focused on the girls, he pulled himself along the Jeep by hand so that he wouldn't pitch forward at the last moment and make a fool of himself. He could tell from the way the girls' faces lit up as they saw him move past the side window, toward the driver, that even though he had never met them before, they knew exactly who he was.

Chip rolled down his window and gave Thad a vigorous handshake before introducing him to the ladies. The blonde was named Rachel, a physical engineer from South Carolina who had just begun her first tour, working with submersible, radio-controlled exploration vehicles. Thad figured that sooner or later, he'd probably see her on the deck of the NBL.

The other girl, the petite brunette, was named Rebecca. Chad introduced her as an up-and-coming biologist. She was only twenty, and one week into her first tour—but already she had impressed Bob Musgrove and the other heads of the co-op program into letting her run her own plant-life photosynthesis experiment. Plant growth in zero gravity was one of the more important areas of study at NASA, now that Mars was in its headlights; creating a sustainable environment would one day involve the secret world of plant biology.

The fact that Rebecca was already being described as a brilliant scientist barely registered; now that Thad was close enough to really see her through the open window of the Jeep, he was having an almost vascular reaction.

Physically, she was stunning. Her hair was jet black, framing a face that looked as if it had been carved from polished porcelain. Her cheekbones were unnervingly high, and her playful blue eyes lit up in a way that reminded Thad of the bioluminescent algae they were on their way to see. She was wearing a white T-shirt and extremely short shorts; even from a glance, it was easy to discern her tight, athletic form. The sliver of bare skin between her shirt and shorts sent chills down his spine, and he actually found himself turning his eyes away. To his utter surprise, he was intimidated by this ninety-pound girl.

He hadn't spoken one word to her—and yet he found himself terrified that she was looking right through him, right past the exciting persona he had created, right through to his core. And he wanted nothing more than for her to react to that core, the way he was reacting to her presence. He felt numb all over. Like he'd been down in the NBL a minute too long—and now he needed the doctor to send them straight to the sci-fi showers.

So he did the only thing he could think of. He completely ignored her, focusing instead on his buddy in the front seat. He made small talk for a few minutes and then quickly hurried back down the line of cars toward his Toyota. His heart was pounding, and he no longer noticed the way the flip-flops grabbed at the jutted floor. He fought the urge to look back over his shoulder, to see if she was watching him. He had a feeling that if he did, there was a good chance he would lose his balance entirely and end up underneath one of the cars. He was completely at a loss to explain the way he felt. If it had something to do with the continuing demise of his relationship with Sonya, well—the timing couldn't have been better. He was in need of a new experience.

As he reached his car, he had a feeling that the new co-ops weren't the only ones who were about to embark on a life-changing adventure.

The long hours are still flowering. I watch them with my eyes closed and remember how I once played a part in a fairy tale. How the most beautiful young woman simply materialized out of my dreams. I was intimidated by her delicate grace, infatuated with her mind, and mesmerized by her body. In the peak of my confidence I was absolutely helpless. French words part her lips. She is thrilled to show me a lichen, she wants to go flying, she brings the heavens to me, she jumps into the water in her little black bikini—first!

22

It happened so fast, Thad never had a chance to react.

He was halfway into his speech, arms wide and outstretched, palms open like he was an actor taking part in some second-rate, drama-school production—Julius Caesar imploring the Senate—except in this case Caesar was naked but for a bright yellow Ocean Pacific bathing suit that clung to his muscular thighs, and the Senate was made up of a bunch of terrified-looking nineteen- and twenty-year-olds huddled together, backs against a rigid outcrop of granite-streaked rock. Thad's own bare heels hung an inch out over the edge of the cliff formation where they were all standing; a mere few feet of mostly level stone that ended in what could only be described as a true precipice; a sheer fifty-foot, stomach-turning drop into the natural reservoir far below. The very magnitude of the drop made what Thad was in the midst of telling the gathered college kids seem all the more preposterous. And it didn't even help that he'd practiced the rallying speech a half-dozen times the night before.

"Okay, guys, here's the deal. I know from where you're standing, this looks pretty intense. And I'm not going to lie to you—the drop behind me, well, it *is* pretty intense. Fifteen, maybe thirty seconds that will feel like a lifetime, and nobody's going to be holding your hand on the way down. Each one of you, the air whizzing by, the water racing toward you—well, you're right to be terrified."

It wasn't *Braveheart* or *Gladiator* or even *Spartacus*—but Thad could see he certainly had their attention. Wide-eyed, barely breathing, the college kids were hanging on his every word.

"Now," he continued, his voice as calming as possible. "Nobody's going to force you to go over this edge. Just like nobody forced you to follow me up here in the first place. Nobody held your hand and told you that this was something you had to do. And it's not something that will end up on your résumé. It's certainly not something that's going to impress your parents over Christmas break. If you do this, if you have the courage to do this—you'll only be doing it for the experience itself. It's not going to help you get into graduate school, or help you get that job offer that we're all hoping to get one day—it's just a single experience, plain and simple. But it will be *your* experience, and that's something no one will ever be able to take—"

And right then, it happened. Without warning, there was a sudden flash of motion from somewhere deep in the crowd of terrified students. And then one of them was hurtling forward, bare legs churning as she drove straight for the ledge. Thad's mouth was wide open as he watched her go, more apparition than twenty-year-old college coed, her petite, tight body somehow 80 percent legs. In that brief second, he realized two things: the girl was wearing nothing but a tiny black string bikini. And he knew exactly who she was.

But before he could even shout out her name, she was rocketing past him. Her body passed within mere inches of where he was standing, his back still to the precipice—and the wind of her rocked him back on his heels. He was barely able to steady himself, his arms windmilling in the effort to reset his center of gravity. He caught a sharp whiff of her floral perfume, the briefest taste of her citrus shampoo from an errant strand of her short hair—and then she was gone, over the edge.

He turned and gaped after her—or more accurately, at the place where she had last been. He could hear the other college kids whispering, awestruck, behind him, but their words barely registered. Staring at the swirl of air and granite dust where she had just gone over, Thad knew that something significant had just taken place. Because in that moment—watching this beautiful young thing throw herself with pure and reckless abandon over the edge—he knew he had found something much more powerful than a simple motivation.

Thad Roberts had just found his catalyst.

In his head—a countdown had begun.

And just like that, without another word to the college kids behind him, Thad threw himself over the precipice after her.

23

Thad's fingers trembled as he dialed the phone number for the third time, determined, this time, to make it all the way to the final digit. He hadn't been this nervous making a phone call since his interview with Bob Musgrove, and though this time he didn't have a picture of the person whom he was calling affixed to the wall above the small desk in his Clear Lake apartment, he could have sketched every inch of her porcelain face, just by closing his eyes.

As he worked his way through the numbers, he thought back to the first moment he had really looked into her eyes: only moments after he had followed her over the precipice's edge, he had burst up through the glassy surface of the reservoir, gasping for air, wildly scanning back and forth to look for her—and there she was. Almost right on top of him, grinning and laughing and splashing water at him like the fifty-foot drop they had both just endured had been little more than a single step. As one by one the other adventurous co-ops followed them over the cliff, Thad spent the time with her in the water, doing his best to make a connection she would remember.

Despite the bond he had felt the minute he had watched her leap over the edge, he found himself slipping back toward his shy old self. That, alone, terrified him; he couldn't let her see past the facade he had created at NASA, the personality he had fought hard to become. But just moments into his first conversation with her, he could tell that he didn't have to worry. She was as swept up in his reputation as everyone else at the JSC. The reason she had jumped, she told him—her cheeks still flushed from the experience—was that she wanted to be more like him. She had heard that he was a person who was good at everything, and this was incredibly compelling to her, even though she couldn't explain to him why.

Somewhere before they crawled out of the reservoir and made their way toward the campfires to dry off, she had given him her number. It had been like rocket fuel in his bathing-suit pocket all the way home to the JSC.

Still, actually calling her had been much more difficult than following her over the cliff. It had been a different phone call that had pushed him into finally taking that next step.

He hadn't talked to Sandra at all while he was home in Utah—but now that he was back at NASA, he had been in almost constant communication with the freckled, mousy girl he'd gone skinny-dipping with, via phone and e-mail, and they quickly became good friends. Though she was still back at school, Sandra had become his confidante, now that he was into his third tour. Mostly, they talked about his problems with Sonya, his growing realization that his relationship with his wife was reaching an end. He certainly hadn't opened up to Sandra about the other thoughts rolling around in his head—he didn't want to get her involved, even with something as innocuous as a mental game. But meeting Rebecca had fallen under the rubric of a confidante's job description.

Thad had expected Sandra to talk him out of doing anything that would put the final nails in the coffin of his relationship with Sonya, but to his surprise, she had been all for it. Maybe because she only knew the Sonya that he'd described to her—the model who spent her time in nightclubs and at the beauty salon, not the outdoorsy soul mate who had rescued him from his troubled childhood. In any event, Sandra had thought Rebecca sounded exactly like the sort of adventure Thad needed.

As Thad reached the final digit, his nerves almost got the better of him, but then he thought back to that moment when Rebecca disappeared over the precipice—and he knew he was making the right decision. He held his breath through three rings, and then Rebecca's voice filled his ears.

Right from the start, she was talking so fast that her words ran into each other. He never even officially got the chance to ask her out, because she'd already assumed that's why he had called.

"I'm shopping right now," she said, over the sound of traffic. "But I should be home at six. You can come over then and hang out with me."

This was going pretty well from the start. Thad hadn't intended to ask her out in the middle of the week—he'd been thinking more along the lines of the weekend—but hell, this was even better. It was a Tuesday, so he had his volunteer firefighting, which meant he wouldn't normally be free before eight or nine—but tonight he was going to make an exception.

At exactly six that evening, he was standing outside the front door to her apartment, just a few blocks from his own place in Clear Lake. She opened the door before he had a chance to knock, another good sign, because she'd probably been watching him the whole time as he'd paced back and forth, working up the nerve to approach the door. She looked fantastic, her white tank top and jeans shorts somehow almost as revealing as the black string bikini from the cliff dive. A portable phone was resting in the crook of her neck, and she cupped her hand over the receiver as she waved him in.

"My mother," she mouthed. "She's giving me advice about guys."

Thad grinned, then leaned forward and gave her a kiss on the cheek. It was something he had picked up during a paleontological dig, which had been run by a team of Canadians. Rebecca flushed at the contact, and he got another wonderful whiff of her floral perfume.

"Tell her I agree; most of us are bad news. But there are a few exceptions."

Rebecca pointed him down the short entryway, which opened up into a small, rectangular living room. The room was completely devoid of furniture, just hardwood floors and bare white walls. Thad raised his eyebrows, and Rebecca gestured toward a spot near one of the windows. He saw a bottle of wine and a pair of long-stemmed glasses.

Rebecca stepped through an adjacent doorway, continuing her phone call, so Thad made his way across the bare living room on his own. He assumed the wine bottle and glasses were obvious enough; he didn't have to be a rocket scientist to figure out where he was supposed to go. He lowered himself to the hardwood floor, sitting cross-legged, his back against the wall. He had a nice view through the window of a small fenced-in yard. Someone had planted flowers along the edges of the grass, and he imagined that it was Rebecca who had been out there on her knees in the dirt, lovingly digging holes for the seeds.

Thad didn't help himself to the wine. He had recovered enough from his childhood not to believe that a glass of wine was a straight ticket to hell, but he still had never been able to shake the feeling that alcohol was some sort of big deal. He liked the fact that for Rebecca, it was something casual and nonchalant and unimportant enough to just be left out in the corner of her living room. For her, having a drink wasn't something she did because it would impress people or because it was rebellious or because it said something about her character. It was just a bottle of wine.

She finally came out of the other room, placing the phone back on its base, and then sat down next to Thad on the floor.

"As you can see, my furniture hasn't arrived yet. So close your eyes and picture a couch, a love seat, and a coffee table."

"I'm not sure I approve of your color scheme. But I dig all that leather."

Rebecca laughed. She grabbed the bottle of wine, popped the cork, and filled both glasses.

"I don't really drink," Thad said, before he could stop himself. He felt like an idiot, telling a hot girl not to serve alcohol. But there was something about Rebecca that made it difficult for him to censor his thoughts. That was particularly scary, considering how much of himself he liked to hide from the people he knew.

"Well, you can take just a sip," she responded. "I'll take care of what you leave behind. Or I can save the rest for later."

Yes, this was going very well, indeed.

...

"Kingdom—*Animalia*. Phylum—*Chordata*. Family—*Balistidae*. And of course the species—*R. aculeatus.*"

Thad leaned back in his chair, one hand absently picking at his conch fritters, as he watched Rebecca. She was halfway out of her seat, leaning so close to the thick glass that ran across the entire wall behind their corner table that Thad could make out her reflection quite clearly, even from a few feet away. In front of her, on the other side of the glass, the colorful, triangular-shaped fish seemed as transfixed as Thad; the pretty little creature was frozen in the water, its fins buzzing like the wings of a hummingbird.

"Shoot," Thad said, still focused on her reflection, "and I thought it was just a fish."

"That's what it wants you to think. Look at its oblong little nose, its sad old eyes. It's not a fish you even think twice about. And then you see its body, the pretty colors and the stripes and even a few errant polka dots, and you start to say okay, maybe it's pretty, but it's just this little fish. And so you swim right up next to it—"

Suddenly she wheeled toward him, leaning all the way over the table between them—and grabbed one of the conch fritters off his plate.

"And then it's suddenly all over you, little fangs tearing out chunks of your skin. The triggerfish is one of the most territorial marine monsters out there. Dive near one when it's protecting its eggs—and look out!"

She took a vicious bite out of the conch fritter, grinning as she chewed. Her energy was so amazing, almost on par with Thad's. There hadn't been a single quiet moment during the drive over to the wharf, and dinner had been one story after another. She shared many of his interests—diving, languages, a love for science—but she was much more than a dilettante, incredibly smart and fast for a girl her age. Walking through the aquarium-walled restaurant—fittingly called Aquarius—was like getting a tour of a marine biology exhibit. She knew the name, phylum, and character of every bit of life behind the glass, and she was not shy about showing off her knowledge. Thad found her incredibly refreshing.

To some degree, she was putting on a show for him. A number of times during the evening, she told Thad that the new co-ops talked about him—about his reputation of being this sort of James Bond type of character. They said he was good at everything, a natural leader.

"Is it true that you snuck onto the space shuttle?" she asked at one point, lowering her voice.

Thad laughed at the idea. He considered spinning it the way his mind often spun things—but with her, he had a very hard time being less than open.

"Close, I snuck into the simulator. But it sure as hell felt like the real thing."

"I can only imagine. Getting that close to actually being in space. I don't think I'd sleep for a week after that."

There it was, that incredible enthusiasm. She felt as strongly about becoming an astronaut as he did. If the aquarium was any indication, Thad believed that she would no doubt impress her way into a position at the JSC by the time she'd finished her three tours. The manner in which she could ring off not only the marine creatures' names and characteristics—but also where they were from, and even how they interacted with each other—she had to have a near-photographic memory, like Thad himself. She was the kind of girl you could only meet in a place like NASA.

Thad waited until they were done with dinner—had left the aquarium restaurant for a stroll down the wharf, toward the huge Ferris wheel that dominated one end of the kitschy boardwalk—to finally bring up what he considered to be the elephant on the pier. His marriage, and the strife that he was going through in his relationship. But Rebecca quickly let him know that she didn't really care about the things that weren't in the here and now. In the here and now, they were two young astronaut trainees, walking along the wharf, talking about space shuttles and triggerfish.

As they approached the Ferris wheel, Thad couldn't help himself; he awkwardly reached for her hand. The minute he did so, he became very nervous—and even worse, she made a weird motion—but then he realized she wasn't pulling away. She took his hand and put it around her waist.

From that moment on, his nervousness was gone. Thad was entirely into her, into her intellect and her personality, and feeling that passionate way that he had first felt when he saw her racing toward the precipice, clad only in that tiny string bikini.

After three rides on the Ferris wheel, rising up together high into the sky over the wharf, so high they could point at the stars and joke about how they would one day be racing between them—her attending to her space plants while he walked around the outside of the spaceship in an EMU—they finally made their way back to her apartment.

Rebecca was only the second girl Thad had ever kissed, and the moment was everything he could have imagined it would be. A little awkward, a little clumsy, his hands not quite knowing where they were supposed to be, the hardwood floor digging into his knees as he leaned forward into her, her back pressed against the wall beneath the window, where she could look out on her little foolish flower garden—the garden that Thad would now protect like a triggerfish hovering over its eggs.

And then the kiss became much more, Thad's hands sliding beneath the tight tank top, his fingers moving up her warm flesh, feeling the ridges of her rib cage and the small swell of her perfect, perky breasts. Her own hands seem to linger just as long on his body, her nails so incredibly delicate against his muscular legs and arms. The heat was rising fast between them, and somewhere along the way Thad thought about telling her to wait, slowing things down so they could talk more and think more, but then she turned to the side, showing off the naked curves of her back. Halfway down her left thigh, Thad saw that she had a tattoo, a little Chinese character. He had been studying Chinese over the course of his three tours, but he didn't quite recognize the character. Rebecca noticed his attention, and smiled at him, cupping her breasts as she twisted to show off the tattoo a little better.

"It means freedom."

Thad felt a tremble move through his body. He knew then and there that if he waited another moment, he was going to open up to her about the one thing he hadn't yet told her about. The secret that he still kept from everyone. *Freedom—to tell her anything. To tell her everything.* But he knew that if he started to tell her, if he described the e-mails and what he was thinking—the mental game he had been playing for so long now—there would be no going back.

So instead, he leaned deeper into her, pressing his mouth against the small of her back and down her thighs, letting his tongue dance against the Chinese character.

As his fingers moved around to the front, slipping beneath the frilly material of her underwear, he knew that he wasn't going to be able to keep the secret from her for very long. Sooner or later, the truth would be as clear as the tattoo on her thigh. And then he would know for sure, if she was really the catalyst he believed she might be, or if she was just another component in another mental game—a fantasy that felt so real he never wanted it to end.

Catalyst or fantasy, he knew for certain: this girl was going to change his life.

People tell me that it wasn't real—that the quiet moments are to be avoided, not enjoyed. But if I can't enjoy the song you once played for me, then I am not defined. It was a harmony that makes me look to the heavens and wonder. It inspires me to seek and explore and to hope for laughter . . . and the rapture of love. Within my collection of permanent echoes the song I remember still plays.

24

The real fun wasn't in the power of the thing—the strength of those massive jet engines, the sheer force of that mechanical monster, built for one purpose only, to lift, to rise, to tear itself free of gravity and physics and sometimes, it seemed, common sense; the real fun came in that moment of sheer helplessness, strapped to a chair, leaning back at a forty-five-degree angle as the beast climbed and climbed and climbed.

And suddenly it wasn't climbing anymore, the great mammoth engines reduced in a whine of reverse thrusters, the nose tipping downward—and then it was falling. The straps came loose and you were out of your seat, just floating in that bizarre way, moving through the padded cabin, bouncing off the walls, the ceiling, the equipment you brought up there to test. Still helpless, but now because the physical laws you've lived with all your life were suddenly gone, replaced by a feeling that was new and unique and wonderful.

Weightlessness. Zero g.

And then the alarm went off, telling you that it was time to strap back in. The craft was now facing downward at a thirty-degree angle, diving at an incredibly high speed back toward Earth, caught again in the grips of gravity and physics. A moment later, the entire sequence began again: the upward climb, the unstrapped moment of bliss, the descent. Again, and again, and again.

NASA had a name for it. They called it the Weightless Wonder, a KC-135 stratotanker known as NASA 931, an airplane that had been specially outfitted for the maneuver. Flying a perfect parabolic route above the Earth, it treated its passengers to as much as twenty-five seconds of weightlessness for every sixty-five seconds of flight. Which didn't sound like much—until you were up there, spinning through the center of the white, cushioned cabin, trying to figure out how to use a screwdriver or plant a tree or maybe even operate a toilet. The ride up was exhilarating enough, but those brief moments when gravity disappeared were another universe altogether. For some people—a full third of those who went up in the thing—it was too much to handle. NASA called it the Weightless Wonder, but everybody else called it the Vomit Comet.

"But it's really a misconception," Sandra was saying, as animated as a cartoon as she bounced around Thad's apartment, using her hands to help paint the picture for him as he lay back against a couch and tried to imagine himself in the scene she was describing. "It's not weightlessness. You're actually falling. Falling around the Earth, at seventeen thousand miles per hour. Altogether you get about twenty minutes of zero g—and it's just amazing."

Her left hand was still making elliptical motions, showing the path of the airplane, and her little freckled face was beaming, as if she had just stepped off the thing. Thad was duly impressed. Sandra was still only nineteen, just an intern, not even a co-op, and she had gotten to do something that he himself had still never done. She had really broken out of her shell, and Thad felt proud that he had been a part of her growth. If it hadn't been for the confidence he had instilled in her over the past year, in person and via their telephone bull sessions, she would never have had the guts to present her project idea to the team in charge of the Vomit Comet. But the fact that they had chosen her work, that was due to her alone; she was a rising star, and no doubt she'd be a co-op by next tour.

"So did you get sick?" Thad had to ask, lacing his hands behind his head as he stared up at the ceiling. In his mind, it was him floating around that cushioned cabin.

"Nah. They give you these pills that make it almost impossible to get nauseous. I think only ten percent of people still have problems. I was too excited to feel sick. And once I started my work, I forgot I was on a plane at all. You know how hard it is to wire a circuit board in zero g?"

Thad could only imagine. He was really happy for Sandra, because she'd had an experience that almost nobody else would ever have. She wasn't even at the JSC full-time, just visiting now and then, but she had created a memory she would carry with her for the rest of her life.

Thad must have gone silent for longer than was appropriate, because before he realized it, Sandra had sidled up next to him on the couch, pushing his legs aside to give herself room to snuggle into the cushions. She was looking at him intensely, and he kept his eyes turned away—because he knew she was about to bring it up again.

"Okay, now you've got to tell me," she started, proving that he had read her correctly. "It's just not fair, you keeping secrets like this. Does it have to do with Rebecca?"

Sandra had been peppering him with questions about Rebecca since their aquarium date had morphed into a full-out love affair; in fact, in the past two weeks, Thad and Rebecca had been inseparable. He had spent every night in her still-furnitureless apartment. They had shared nearly every meal together, had spent the weekends camping, alone in a tent. They had made love every night, woken up naked and entwined together.

He hadn't spoken to Sonya much since meeting Rebecca. She had called a few times in the beginning—but over the past nine days, she had given up trying to reach him, and there was no doubt she suspected that something was going on at the JSC, involving someone else. Thad didn't want to hurt her—but Rebecca had become much more than a fling to him.

He was in love. As always, it was difficult for him to separate what was fantasy and what was real—but the feelings he was having for Rebecca felt like both, fantasy and real. So he had thrown himself into her with total abandon.

And the more time he spent with her, the more the thoughts in his mind had grown clearer, the more the mental game had started to take a more physical shape. In the process, the game had become a secret that he was finding increasingly more difficult to keep. Both Rebecca and Sandra had noticed—especially during moments like this, when he went silent, playing it through in his head, like a movie on a spool that kept running over and over. With Rebecca, he had resisted by telling her that it was something he needed to protect her from, that if she really wanted to know, he would tell her—but that keeping it from her was for her own good. With Sandra, he had simply remained mysterious. But it was obvious from the way she was gripping his calf, her mouse fingers tightening into a claw, that she was getting tired of the subterfuge. If she was really his confidante, she felt she had a right to know.

"Okay, if it's not Rebecca, is it Sonya again? Because I still think you're doing the right thing—"

"Why does it have to be about a girl?"

"Because you're a slut," Sandra responded. Then she grinned. Two girls in one lifetime was about as far from a slut as a guy Thad's age could get. Though he *was* technically married, and sleeping with a twenty-year-old beauty. But he no longer saw it that way. He was sleeping with the girl he was in love with.

"Okay, if it's not Rebecca or Sonya, then what is it?"

Thad slowly sat up, crossing his arms against his chest. He looked at Sandra, trying to read the freckles on her cheeks. She really wanted to know—and in truth, he really wanted to tell her. But the minute he said it out loud, to someone here, in the JSC—it was going to become real in a whole different way. Gordon was so out of it and so out there—hell, Thad was pretty sure the stoner still had no real clue about what they were even e-mailing about. Gordon was playing a game, too, though

Thad could never be sure what game the guy thought it was. But Sandra would understand—she would think it was impossible, because it was, but she would understand. And even just knowing about it—that would make her part of the scheme. Thad didn't want to be responsible for that. He had helped her come out of her shell—he didn't want to do something that could be detrimental in even a small way.

Still, the idea of talking about it—even in a gentle way—was appealing. He decided that it couldn't hurt to at least feel it out, without giving away anything important.

"It's not so much of a secret, actually, as it is a hypothetical."

"Like, hypothetically, whether or not you believe in love at first sight? Whether someone can fall so deeply in love in a couple of weeks—"

"It's not about love. It's more of a moral hypothetical. Let's say you were in a situation where you knew that there was somebody who owns something that's clearly theirs—yet they throw it in the trash, they identify it as trash. And let's say you had the opportunity to grab this thing before anybody knew. And even though *they* had labeled it as trash—you could sell it for a lot of money."

Sandra was watching him carefully, her left hand still resting gently against his calf, but her claws had retracted.

"A lot of money," he repeated. "Would it be morally all right to take this thing and sell it?"

Sandra's eyes never left his face.

"What are you getting at?"

"It's a hypothetical."

"Thad—"

"Just go along with it. I really want to know your opinion."

"Okay, hypothetically, I think it would probably be okay. Since there's no harm being done, because whoever owned the thing has already deemed it trash. You're kind of creating value. So in a way, it's actually a good thing."

Thad was getting warm inside, like when he'd taken a sip of Rebecca's wine before letting her finish the glass.

"Now this is torture," Sandra grumbled. "You know you can trust me. I mean, I've known you like ten times longer than Rebecca, and don't forget—I saw you naked first."

"It's not a matter of trust. It's just . . . it's something pretty crazy. And it's a lot safer if you don't know."

"Now you've really got to tell me. I'm not scared. I don't get scared anymore."

Thad laughed. He really didn't want to tell her, but he was running out of excuses. Just like with Rebecca—it was doubly hard to keep a secret that you didn't really want to keep. And was it really anything more than the hypothetical he had just brought up? Wasn't it still just a hypothetical heist?

"I'm going to give you only one chance," he said finally. "A little game. If you win, I'll tell you. But if you lose, you can never ask me again."

"What sort of game?"

He reached over the arm of the couch and retrieved a little cardboard box from the floor. Inside the box was a stack of flash cards. Each had a Chinese character written on one side, an English translation on the other. Thad had gone through them many times in the course of his Chinese lessons, and even so, he still found them difficult. Reading those twists of black ink was as hard as intuiting an expression from a matrix of freckles.

"I'm going to show you twenty of these flash cards and tell you what they mean. Then I'm going to shuffle them and show them to you again, one at a time. If you get all twenty right, I'll tell you what you want to know. And if not—"

"I can never ask you again."

She shifted her body so that she was facing directly toward him, little hands on her lap, her face a mask of concentration. Almost immediately, Thad regretted offering up the game. Still, twenty characters? She couldn't possibly get them all right.

"Here we go."

He held up the first card, showing her the convoluted twists of ink that made up one of the more recognizable Chinese words.

"This one means 'love.' I guess it's as good a place to start as any."

"At the very least," Sandra joked as her eyes flicked back and forth over the flash card, putting it to memory, "I'm going to have some great ideas for a tattoo by the end of this."

Thad sighed, wishing he hadn't told her so many details about his time with Rebecca. He held up the next card, showing her another character.

"'Umbrella.' Not quite as popular as tattoos go, I imagine."

And on and on they went, through the flash cards. Thad didn't move too quickly, but his pace wasn't slow either. Within a few minutes, he had been through all twenty, and then he began shuffling. Sandra barely seemed to be watching him, but he could tell she was going through the cards over and over again in her mind.

Carefully, he began showing her the shuffled cards, one at a time. By the fifteenth card, he felt his cheeks flushing red. He had underestimated her. Her memory was even better than his own. As he reached the twentieth card, his fingers were shaking. He held the card up in front of her—and she paused only a moment. Then her face broke out in a huge, freckled grin.

"'Happiness,'" she nearly shouted, her voice bouncing off the walls.

Shit. Thad thought about ignoring the results, simply telling her again that he just couldn't risk getting her involved. Really, it was for her own good. But she had played the game, and won.

He leaned close, and lowered his voice.

"Okay," he said—and then he started talking.

I haven't stopped loving you, Rebecca, but I have accepted our separate paths. I hope that someday you allow me the closure I have longed for, that you forgive me for not being there forever, for taking a foolish risk that jeopardized our union. Perhaps you desire not to be friends, perhaps you have succeeded in convincing yourself that my love was not genuine. I hope these things have made the past few years easier, but as the wound heals I hope you find it in you to share your mind with me.

25

Rebecca ricocheted through the compact, galley-style kitchen, first tossing her purse on top of a pile of unopened mail, then grabbing a tied-off plastic bag full of recyclables with one hand while opening the refrigerator and retrieving a pair of long-neck bottles of root beer from the fridge door with the other. Still moving—hell, the girl never stopped moving—she offered one of the bottles to Thad, who was half skipping behind her, trying desperately to keep up. Then she yanked open the sliding-glass door that led to the small balcony where she kept her garbage. The bag of recyclables landed with a clunk next to an overstuffed garbage can, and Thad had the feeling that his girlfriend was taking care of the recycling for her entire building.

"It's just so crazy what this place used to be like," she gushed as she spun back through the kitchen, using her arms to hoist herself onto the edge of the counter, her slim bare legs crossed at the ankle. "I mean, it's still cool now—but back then it was just insane. These guys, these cowboys—they were basically strapping themselves onto the tips of missiles. Blasting off into space, trying to get one foot onto the surface of the moon—actually competing for the opportunity to go on what was basically a suicide mission—and all of it taking place in a time when their biggest supercomputers were less sophisticated than my cell phone."

Thad laughed, but he was no less awed by the thought than she was.

He guessed that conversations like this were taking place all over the JSC; tonight had been the annual co-op ritual where everyone gathered together to watch the movie *Apollo 13*—the story of one of the ill-fated attempts to duplicate Neil Armstrong's walk on the moon.

Having spent time in the old Mission Control room, where the events documented in the movie had taken place, Thad had seen for himself how rudimentary some of the technology had been during the Apollo era. He'd sat in the actual flight director's chair, his fingers touching the very consoles that had been used in those missions. But nothing about seeing the original Mission Control made him feel superior—quite the opposite, seeing what those men had to work with, the truly historic level of bravery—it only made him feel utterly small.

"Mars isn't going to be all that different," Thad responded, putting his hands on her bare knees as he leaned in, planting a kiss on her lips. "We're still going to be strapping ourselves into a tin can attached to a missile. The toy's a lot shinier—but the project is going to be just as dangerous. It's going to take a special type of person to embark on what might end up a suicide mission. Someone willing to take a chance, to make a leap of faith."

Rebecca put her hands on his shoulders, feeling his muscles through his shirt.

"A leap of faith; I like that. Like, maybe, finding yourself madly in love with someone you've only known for ten days."

She was grinning, but Thad couldn't really read her expression. He wasn't sure whether she was talking about herself or about him. They had been using words like *love* and *forever* since their very first evening together, but it was hard to know whether those sentiments were just symptoms of her youth, or symbols of his passion; Thad only knew for sure what he was feeling. Which was beyond anything he could remember ever feeling before. He'd always loved Sonya, but he didn't remember it ever being this all-encompassing, mind-bending thing.

He realized that he had once again slipped into that other place, going silent as he stared right through her. Her grin had turned down at the corners as she watched him, her hands going limp against his chest.

"There it is again," she said. "That thing you do, sometimes right in the middle of a sentence. Sometimes even when we're making love. I know there's something you're keeping from me."

Thad stepped back, taking his hands off her knees. It wasn't the first time they'd had this conversation. He'd explained again and again that it wasn't some other woman, some relationship, or anything to do with Sonya. But now they were at a point where she was asking about his secret almost every time they were together.

"It's not that I don't want to tell you. It's just that, well—I'm thinking about doing something that's technically illegal. I mean I'm pretty sure I'm not going to do it, but even talking about it feels like it could be dangerous."

He felt his adrenaline rising, because this was the closest he had come to telling her. And he knew he was standing on the top of a slippery slope. Telling Sandra hadn't made the secret any easier to keep; in fact, he'd gotten such a rush out of talking through his plan with her, he was having a hard time not shouting it from the rooftops. And he could see by the way Rebecca's eyes had gone really intense that she wasn't going to be content with another excuse.

Maybe it was time to tell her. Meeting her had pushed him forward in the mental game; that very morning, he'd sent another e-mail to Gordon, asking him to research the sister-in-law of the Belgian rock hound, the woman named Lynn Briley, because, at least via e-mail, they were coming close to actually setting up a face-to-face meeting. Not just any meeting—an exchange, goods for cash—as if it were really that simple, as if there weren't a step in between the e-mails and handing over the parcel in exchange for a hundred thousand dollars in a suitcase. A step that was still entirely fantasy, entirely impossible.

Thad breathed deeply—and then, he just let it out. It was like he was back on that cliff, heels hanging out over the drop—but this time, it was he who was going to jump first.

"I have this idea. It's completely insane. And it's also impossible. I'm thinking about stealing a safe full of moon rocks. It's in an impenetrable lab, protected by the highest level of NASA security. The samples are considered trash because they've already been worked with and experimented on—but they're incredibly valuable. I've already got someone who wants to pay me a hundred thousand dollars for a little piece of the moon."

Rebecca was still staring at him, her eyes wide and her lips parted so that he could see just the tips of her teeth. Even as he was talking, he was thinking it through, not just how elaborate and ridiculous and impossible the actual heist would be, but internally, he was asking himself why he was even still playing this game, why he didn't just erase all the e-mails, lose the contact info for the woman in Philadelphia, maybe even throw out Gordon's phone number—just forget about the whole stupid thing. And yet he kept talking.

"I mean, a hundred thousand dollars, it's a hell of a lot of money. The things you and I could do with that money—we could go to Africa, and you could study the plant life there. We could put the money toward starting our own lab, so we wouldn't need to compete for a grant or wait until we were old enough. We could start right away, doing all the things that we've talked about doing. But the money, it's only part of it."

He kept expecting her to interrupt. He fully expected her to shake her head, glare at him like he was crazy, talk him out of it. He expected her to tell him that it sounded exciting, but of course he shouldn't do it, that he would be risking everything, that he would get in huge trouble, that it was a really bad idea. But still she remained silent, letting him finish the thought that had been building since the moment he'd first laid eyes on her.

"Rebecca, I want to give you the moon. I mean, a piece of the moon. Like the astronauts we just watched in that movie, the cowboys who took that crazy chance, all to set foot where only a couple of people have ever been—I want to give you that. I want to give you the moon."

Thad realized that his eyes were watering. It sounded so crazy, so stupid, and—well, he didn't really know how it sounded. But he did know that he actually meant it.

The kitchen was dead silent, the scene frozen like a photo in an album. Then Rebecca's eyes lit up, and she was grinning.

"That sounds so romantic. Let's do it."

And in that instant, Thad knew that he'd been correct; Rebecca was his catalyst. His instant, passionate, consuming love for her had shattered the glass wall in his mind that separated fantasy from reality. The fracturing that had begun long ago was now complete, and the mental game he had been playing had gone from a thought experiment to a project, no different from any of the projects he had worked on at NASA, no less real than the Space Shuttle Simulator or the space station that was sunk into that six-million-gallon pool.

Without another word, Thad leaned forward and pressed his lips against hers. Slowly at first, then gaining in intensity. To Rebecca, he was everything he'd ever wanted to be: exciting, adventurous, James Bond. He didn't know if anyone had ever promised her the moon before—but he was the one guy who was going to deliver.

She was his catalyst.

And now it was only a matter of time.

26

Nothing got the old heart pumping like the shrill, piercing wail of a telephone cutting through the dead, still heat of a summer morning. It wasn't particularly early, but Axel had been dozing pretty deeply, his rounded form splayed out comfortably across the small couch that ran along one wall of his living room. The TV was still on a few feet away, tuned to the French murder mystery he had been watching when he'd first closed his eyes for a moment—but one more metallic ring reverberating through his head, and he knew for sure that the sound wasn't coming from some faraway sound studio in Paris. It was echoing off the walls of his own home in a quiet corner of Antwerp.

It had been a perfect weekend morning before the sound of the ringer had ruined it; perfect, because the kids were locked up in the kitchen frantically studying for their exams, and because Christel was out having breakfast with a friend. Which meant that Axel was able to enjoy some quality time with his favorite couch cushions. Since he still hadn't been sleeping that well at night—his mind locked into the drama he imagined was unfolding far across the ocean—the minutes alone with the couch were as valuable as polished topaz.

Ten days without any contact from either the FBI or Orb Robinson had certainly taken its toll on Axel's psyche. It was kind of like watching the French murder mystery, but with the sound off. He could only fantasize about what was really going on. For all he knew, the whole thing had fizzled and disappeared. The hoaxer might have finally grown bored with the game, moved on to something else. Maybe he was now sending out e-mails, posing as a Nigerian banker, or the cousin of a deposed prince. *Just send a cashier's check, and my fortune will be yours.*

But as soon as Axel heard his son, Sven, answer the phone through the door that separated his living room from the kitchen, as soon as he registered the shocked tone of the fifteen-year-old's voice, he had a feeling that his wait was suddenly over. He sat straight up, shaking the last vestiges of sleep out of his eyes, just in time to see his son stick his head out through the kitchen door, the phone cupped against his chest.

"Dad, I think it's for you. It's an American."

Sven looked like he had seen a monster, and that got Axel's heart pumping even faster. He indicated with his hand that he was going to pick up the receiver in the living room, and that his son should hang up once he was on the line. Then he rose, flattening the wrinkles out of his slacks with his palms, and crossed to the computer desk in the corner of the room. He didn't know why, but for some reason he wanted to look presentable—even though he was only going to be talking over the phone. It wasn't often that he got calls from America. Actually, it wasn't ever.

He cleared his throat, then picked up the receiver.

"This is Axel Emmermann."

The American on the other end of the line quickly introduced himself as Special Agent Nick Nance of the FBI. Axel felt his shoulders pulling back, his chest sticking out as he heard the words. E-mails were one thing, but now he was talking to a real-life FBI agent. His superhero status was quickly rising.

Very rapidly, the official-sounding man on the other end of the line brought him up-to-date. Even though Axel hadn't heard anything for the past week and a half, it turned out that the FBI had been quite busy. Agents posing as Axel's brother and sister-in-law had continued to lead

Orb Robinson along, getting him to the point where they seemed actually ready to enact an exchange. They were in the process of setting up a face-to-face meeting. Robinson still didn't seem to have the actual items in his possession, but he was moving forward as if he could get them at any moment.

Agent Nance explained that "Lynn and Kurt" had confirmed receipt of the hundred thousand dollars, and had e-mailed Robinson, telling him that they trusted him, that they believed his claims were truthful and were ready to buy what he was selling.

Axel had to fight the urge to start jumping around the living room. The French murder movie seemed like such a trifle now, compared with the real mystery that he was an integral part of. He was actually talking to the FBI, and they were going to meet with this hoaxer. He couldn't wait until his wife got home so he could tell her what was about to happen. And then Nance added something to the conversation—something Christel wouldn't find quite as enthralling.

"Now, there's a chance this Robinson might try and call you directly. I don't think it would be that hard for him to find out where you live, and get your phone number. So we're thinking about installing a recording device so if this happens, we can listen in."

Axel swallowed, focusing on the comment that Robinson wouldn't have much trouble figuring out where he lived. He immediately pictured his kids in the kitchen, huddled over their schoolbooks. It was a terrifying thought. Certainly, this bit of information he would leave out of the upcoming conversation with Christel.

"And if this actually goes down," Agent Nance continued, "if we do arrest this Orb Robinson—we need to ask—would you be willing to testify? We'd bring you here to the U.S. and put you up in a hotel for the length of the trial, if we deemed it was necessary. Is this something that you would be willing to do?"

Hearing this, Axel had to sit down in the chair in front of his computer. That he could be asked to take part in bringing this criminal to justice—not just being the middleman in an e-mail investigation, but actually taking physical part, becoming a player in the drama—wow.

"I would be honored to take part in your judicial system."

Axel Emmermann the superhero, becoming Emmermann the star witness. It certainly would beat an afternoon at the popinjay field.

But sitting in the chair—looking at the computer where this had all started—Axel began to have a thought. The way Nance was talking, it was beginning to sound like this might somehow be a little more than a hoax. If they were thinking of bringing Axel all the way to America . . . well, it wouldn't be because someone was trying simply to make money on the Internet.

"Special Agent Nance, are you beginning to suspect that this Robinson might be trying to sell authentic moon rocks?"

There was a long pause. For a brief moment, Axel could hear the buzz of the international phone line.

And then: "It's not impossible."

With that, the FBI agent thanked him again for his time and then gratefully hung up. As Axel replaced the receiver, the words continued to reverberate through his mind. *It's not impossible.*

Christ; what, exactly, had he stumbled into?

* * *

Axel was still sitting in front of the computer—mulling over what he had just learned, waiting for the sounds of his wife's heels on the front steps so that he could relay the developments he'd just learned of, and sure, brag a little bit about the possibility that he could soon be racing halfway around the world to bring a master criminal to justice—when an icon appeared on his computer screen indicating that he had a new e-mail. One click later, and he saw that it was once again from the FBI, the same Special Agent Nance:

> Mr. Emmermann.
>
> It was nice to talk to you this morning. I neglected to ask you for

> your help in putting together some questions that should be asked by Lynn. Since my knowledge of lunar materials is limited at best, I was hoping you could provide questions to via e-mail that will lend to my/our credibility. Any help would be greatly appreciated . . .

At first, Axel was quite puzzled by this new e-mail, which was accompanied by an even longer explanation of what Nance was looking for. It seemed the FBI was asking for Axel's help in explaining how their agent could best recognize real moon rocks—and furthermore, how she would be able to tell the difference between moon rocks that had actually come, by hand, from the moon, and ones that had fallen to Earth as meteorites. Wouldn't the FBI have their own specialists who could assuredly do a better job of explaining this than an amateur rock collector such as himself?

But as Axel worked it out in his head, he realized that the FBI's request made sense. Orb Robinson had written that the moon rocks were not currently in his possession—which meant that he intended to steal them.

There was only one place on Earth from which he would be able to steal the amount of moon rocks he was talking about: the Johnson Space Center in Houston. If the FBI had wanted to talk to specialists who could help identify real moon rocks . . . well, the place they would normally go was also where Robinson's crime would take place—the JSC.

So obviously, the FBI couldn't go there for information; they couldn't yet know who Orb Robinson really was, and had to suspect anyone with access to the Apollo rocks. It was hard for Axel to believe that someone who worked at NASA was planning to steal moon rocks; not just because they were national treasures, but if you were lucky enough to work at NASA—in the same hallowed buildings where the Apollo program had taken place—how could you throw it all away for a hundred thousand dollars?

In any event, Axel was more than happy to continue to help the FBI. After his first contact with Robinson, he had done a fair amount of research into moon rocks. With the help of his notes, he began to compose his response to Agent Nance.

Moon rocks were usually light in weight and color, made up mostly of basalt, with a mix of pyroxene and feldspar within, easily recognizable by a geologist using a magnifying glass. But this information wasn't going to be all that helpful to an agent during a sting operation. Especially an agent posing as a rock collector—and not a professional geologist.

But there was a much simpler way to recognize a moon rock—and especially to distinguish a moon rock that had actually been picked up by hand—by an astronaut on the moon—from a meteorite that might have been stolen from a museum.

As most people were aware, the moon had no atmosphere. Which meant that anything that hit the surface of the moon—from a giant asteroid to a tiny grain of sand—hit the ground somewhere between ten thousand and eight thousand kilometers per hour. On Earth, such objects burned up in the atmosphere because of air friction, but because the moon had no atmosphere, dust and sand were continually raining down to the surface, at these immense speeds.

So any rock from the moon would be covered in tiny impact craters. These craters were called "zap pits," ranging from a few microns in size to as big as a few millimeters. They would be easily recognized, even without a microscope: a tiny black-glass center surrounded by a halo of concentric circles, much like the large craters you could see through a telescope when you looked at the surface of the moon.

As Axel sent the new e-mail off to Agent Nance, part of him wished he could follow that little electronic packet of information around the curve of the Earth. He wished that he could walk into that meeting place, with a suitcase full of cash, and sit down across from this master criminal, this person who would dare to steal a national treasure. He wished that he could look this man in the face and tell him, It was me who brought you down. It was Axel Emmermann who caught you.

And then he remembered how he had felt when Nance had told him that this Robinson could easily figure out where he lived. And he quickly changed his mind.

Axel was the kind of superhero who was happy to bring justice to the world, from the comfort and security of his cozy Antwerp lair.

27

Ten. Nine. Eight . . .

Friday morning, a little after seven A.M., and Thad was moving quickly down the central hallway that bisected the fourth floor of Building 31, counting under his breath as he kept one eye on the deserted territory up ahead and the other pinned to the rotating security camera jutting from a storklike metal strut embedded next to one of the ceiling's fluorescent lighting panels. As he had predicted, so far his progress through the life sciences complex had been uninterrupted; any self-respecting scientist who would show up to work this early in the laid-back atmosphere that dominated life sciences at NASA would either be too new to think twice about seeing a co-op wandering the halls or so caught up in a brain-consuming project, he wouldn't notice Thad at all.

And even if someone cognizant did happen across Thad—in his blue NASA polo shirt and khaki pants—the only unusual thing about his demeanor was that his gait seemed a little off center; in fact, if anyone looked closely, they might have noticed that he was moving so near to one side of the hallway that his right shoulder brushed against the concrete. His face, however, was perfectly calm, his expression muted—even as he suddenly shifted to the other side of the hallway, his left shoulder now kissing concrete.

Another flick of his eyes confirmed what he already knew: he'd now moved out of range of the first rotating ceiling camera and only had to avoid the final one, planted all the way at the far end of the hallway. It, too, had begun its own innocuous arc—filming the area where Thad had just been.

As easy as that, Thad thought to himself. A little dance step, a shuffle to the left, and he was a ghost. Of course, for the moment it was easy to play calm; he wasn't doing anything wrong. He was just walking undetected through the building where he had worked for two semesters as a co-op. If, by some odd twist of fate, he did run into someone he knew, there were a dozen explanations for why he might be there on an early Friday morning. The only people in the world who knew the *real* reason he was back in Building 31 were his two pretty accomplices, his new girlfriend and his confidante—neither one a hundred pounds soaking wet.

Fighting back a smile as he pictured Rebecca and Sandra, both waiting in his apartment for the phone call that would let them know that Phase One was complete, he slowed his pace, finally stopping as he reached a closed door located near the center of the long hallway. Bare inches away, midway up the door's frame, was one of the electronic cipher locks Thad had grown so accustomed to in his years at the JSC. In fact, he had even watched this particular cipher lock be opened a handful of times. He had never gotten close enough to look over anyone's shoulder to even attempt to guess at the five-number combination—but that would have made what he was planning to do way too simple, and now that he was determined to see it through, he relished the idea that nothing was going to be easy. As it was for any good scientist, it was the complicated, sophisticated projects that got his juices flowing. Maybe even more than the money, this was now about the thrill of doing the impossible.

Thad pressed his back against the concrete wall, checking the long hallway again to make sure no one was nearby. Then he quickly reached into his left pocket and retrieved a small plastic makeup compact;

originally, it had been Rebecca's, a shade of blush that really brought out the contrast between her porcelain cheeks and her bright blue eyes. The thing no longer contained blush. When he opened the compact with a flick of his left thumb, the powder inside—a unique concoction of his own creation—glistened a bit in the high fluorescent lighting, and Thad wondered for a moment if he'd gotten the concentration wrong. But when he gently shook the compact, evening the powder out, the glistening abated, and he exhaled. This would work. *This had to work.*

Carefully, he removed a small brush from his other pocket and dabbed it into the powder. Then he began to apply the brush to the keypad of the cipher lock, making sure to completely cover each numbered key with the powdery substance. Leaning close, he blew off the excess powder—then stepped back a few inches to survey his work. Even from just a few feet away, there was no real visible trace of what he had just done. Satisfied, he closed the compact and jammed it back into his pocket, along with the little brush. Then he calmly continued down the hallway. As he reached the next corner—passing right beneath the rotating security camera—he fought the urge to glance back one last time at his handiwork. Under his breath, he was no longer counting off the seconds; instead, he was humming to himself—the theme from the movie *Mission: Impossible*. Earlier, when he'd been alone in his old lab a few floors away, it had been the music from the James Bond franchise that rumbled out of his throat as he carefully mixed the compound—equal parts fluorite, gypsum, and talcum. All had been easy to find in the chemical cabinets at NASA, but even so, he couldn't help but feel like a spy or an action hero as he'd prepared the ingenious concoction. Even the name he and the girls had given this portion of their preparation—*Phase One*—made Thad feel like he was part of something epic, an adventure he'd one day tell his grandchildren about.

Powder in a keypad, Phase One—it really was James Bond kind of shit. But still, he knew that he hadn't yet crossed any real line; he hadn't yet done anything that he couldn't turn back from. Powder on a keypad, a dozen e-mails with a potential buyer—it was still little more than a mental game. But Thad also knew that within forty-eight hours, this would all change. Because he was determined now; the plan was in motion.

Phase One was complete. Which meant it was time for Phase Two.

Seven. Six. Five . . .

. . .

> Orb,
>
> I'll handle this for Axel. He's explained to me a good bit about what we're doing and the need for caution and discretion. It seems to me that you and I are going to have to make arrangements to meet somewhere to make sure we're getting what we think we are. This is a rare opportunity and calls for us to be very careful. When and where are we going to be able to get together? I travel a good bit but will certainly make arrangements to see the merchandise wherever it might be necessary. I understand you are in Tampa, Florida. I certainly wouldn't mind taking a trip to Florida. I look forward to hearing from you.
>
> Lynn.

A throbbing burst of high-octane Christian rock exploded out of the dashboard speakers of Thad's dilapidated, bright green Toyota as he navigated through the South Houston rush-hour traffic. He had one hand on the wheel while the other leafed through the stack of printed-out e-mails that took up much of the empty passenger seat next to him. The Christian rock was more than a little annoying—and by no means his first choice—but the Toyota had made the acoustic decision for him, as its pathetic excuse for a radio had frozen on the one channel Thad would have eagerly tried to avoid. But at the moment, stuck as he was in traffic on his way to Phase Two of the Plan in Motion, anything was preferable to silence. Silent, Thad couldn't think past the bolts of nervous energy that were playing havoc with his internal organs—and at this stage of the game, he needed to be able to focus entirely on the preparation at hand.

At the next red light, he used the few seconds of nonmotion to leaf past the top e-mail on the pile—which happened to be the first real message he'd gotten from the sister-in-law of the Belgian rock collector, Lynn Briley—to the more recent e-mail he'd received from Gordon. Just as Thad had done with Emmermann, he'd asked his Utah buddy—he wondered if the word *accomplice* was now more fitting—to check out the Belgian's American relative, to make sure she was who she said she was. Gordon hadn't found much, but there was at least evidence that the woman existed—and confirmation of a few details of her story:

> Hey, Orb.
>
> Here is the only thing I found on Lynn Briley. She is a publisher in Glenside, Pennsylvania, it seems, and then there's the Web site address. Nothing else for now.
>
> Fractal.

Thad found it slightly amusing that Gordon had begun referring to him by the nickname Gordon himself had created—Orb. Thad thought Gordon's own handle was much more indicative of his pothead friend's disjointed character: Fractal. But it was certainly better and safer to use the nicknames than to use their real names—Thad just wished he could have devised the handles on his own. He didn't like that any element of the scheme—even something as simple as code names—was not of his making. Even worse, the face-to-face meeting with this Lynn Briley—and the exchange of money for moon rocks—was now going to have to take place in Florida, because for some inexplicable reason that's where Gordon had set his fictitious Orb Robinson.

Then again, Florida wasn't the worst choice in the world—it was far enough away from Houston to allay some of Thad's fears, but it was still reachable by car. Thad had no intention of trying to get on an airplane carrying the contraband that he'd soon have in his possession.

Contraband. It was still hard to think of it that way, such a loaded term, like he was going to be dealing in drugs or some other dirty, underworld substance. He knew that the thing he was after was much more precious—even if NASA had labeled it trash. It was the most valuable thing in the world, actually, and even if he was only going to get $100,000 from the woman, it was going to be a heist of historic proportions. And as he engaged in Phase Two of the preparation, Thad had every right to think of himself in historic terms.

Restacking the e-mails on the passenger seat, he took a right at the next intersection, then navigated his way through a patchwork of suburban streets until he came to a driveway he recognized from a handful of previous visits. As he had arranged, the purpose of his visit was parked right next to the curb, leaving just enough room for him to get by; a moment later, he'd parked his Toyota halfway down the driveway. He retrieved the e-mails, shuffling them into a manila folder that was wedged between the two front seats. Taking the folder with him, he stepped out of the Toyota just in time to see Chip come out the front door of the small suburban house. Chip gave the Toyota one look, then rolled his eyes.

"Oh, yeah, this is a great deal. No wonder you couldn't find anyone closer to campus to help you out."

Thad laughed as he tossed Chip his car keys. Then he started toward the Jeep Cherokee that was parked along the curb. It was just as Thad remembered it from the Galveston ferry: almost as scuffed and aged as the Toyota, with mud etched into the tires and a spiderweb of tiny cracks in a corner of the front windshield. But the thing was almost twice as big as the Toyota, and with the backseat down, it was going to be perfect for what he had in mind. Even more important, Thad could easily make out the NASA parking sticker affixed to one of the side windows.

"I promise to return it in just as good condition as it is right now.

And it's only for the weekend. We should have my friend moved into her place by Sunday night, at the latest."

"Take as long as you need," Chip said as he turned back toward his house. "The keys are in the ignition. But I want dibs on the skydiving excursion you've got planned for next month."

"I promise, you'll be the first one out of the plane. Heck, I'll pack your parachute myself."

Thad slid into the front seat of the Jeep, twisted the key, and grinned as the ignition turned over. Again, this felt almost too easy. Chip hadn't suspected anything at all—and why should he? Helping a friend move apartments was a perfectly good reason to need a car as big as the Cherokee. There was only one more component to Phase Two—and then Thad would be able to call the girls to let them know he was moving on to the final phase of the preparation.

It took about ten minutes of driving for Thad to find what he needed next. As he pulled a sharp left into a strip-mall parking lot, he glanced about to make sure there weren't any signs of security or parking-lot cameras. Then he pulled the Jeep to a stop between a pair of American-made cars, near the very back of the lot.

He got out of the Jeep, then quickly crossed to the back of the closer car—a Buick that looked to be at least fifteen years old. Thad bent down behind the rear bumper like he was about to tie his shoe—and then, in one quick motion, slid a small screwdriver out from where it was taped within his sock. Of course, he could have carried the screwdriver in his pocket—but that would have felt much less James Bond.

He rapidly went to work on the Buick's license plate. The first screw gave him a little bit of trouble, and he was sweating by the time he got it free—but the other screws went much easier. Within a few minutes, he had the license plate off and moved back behind the Cherokee. Another five minutes, and he'd removed the Jeep's license plate and replaced it with the Buick's. He tossed Chip's license plate into the rear of the Jeep, then jumped back into the driver's seat. As he reentered traffic, he realized that his heart was beating fast. He still hadn't crossed any real lines—but now he was driving the getaway vehicle. A Jeep that wasn't associated with him, that had a NASA sticker affixed to a window and a stranger's license plate above its rear bumper.

Four. Three. Two. One . . .

. . .

Almost five hours later, Thad was really breathing hard, putting all of his weight into his shoulders, straining the muscles in both legs as he shoved the motel bed, inch by inch, across the vomit-colored carpet. He hadn't expected the damned thing to be so heavy; everything else in the pathetic little motel bedroom looked flimsy as hell, from the color TV bolted to the fake-wood bureau by the door to the light fixtures that hung from the chipped plaster walls. Rebecca had picked the motel, and it was obvious she had chosen it right out of the yellow pages. But despite the horrid decor—of which the vomit rug was only the centerpiece, highlighted by a pair of cheap-looking paintings of hunting dogs above where the bed used to be—the motel was ideal for a couple of reasons. First, it was right off the highway, which meant it wasn't too close to the JSC campus, but it wasn't so far away that they would have to spend hours in transit. And second, the place looked nearly vacant; Thad had counted only three other cars in the parking lot, and he had made sure to pick a room on the first floor, surrounded on one side by the ice machine and on the other by what appeared to be a janitor's closet. With any luck, there would be nobody nearby when they arrived after the heist.

Thad felt a thrill move through him, even as he continued to struggle with the unusually heavy bed, as he repeated the word under his breath. *Heist.* It sounded so cool in his ears. *The heist of the century. The heist of the millennium. The great moon rock heist.*

He laughed out loud, and with a final burst of energy managed to shove the bed the last few feet so that it was finally right up against the wall. Then he stepped back, working the cricks out of his shoulders as he surveyed the room. Now there was plenty of space for what might be necessary.

He crossed to the bureau and retrieved the oversized duffel bag that he'd placed next to the TV. He unzipped the duffel, and first pulled out a pair of folded-up tarplike sheets, which he spread out over the nausea-inducing carpet. Then he returned to the duffel and, one by one, laid out the tools he'd purchased from Home Depot along with the tarp—a pretty wide variety, because he wasn't certain what he was going to need. After the tools, he retrieved a large fishing-tackle box, three pairs of rubber gloves, a notebook, and a folded-up mailing box.

After he'd laid everything out, he stood back, smiling. It was a nice-looking staging area. The tools themselves weren't exactly high-tech; the most sophisticated of them were basically a saw and a handful of industrial-strength blades. But he had been working off a limited budget. And he was proud that he was planning to do this with such meager supplies. It was one thing to pull off a heist like this with the best supplies that money could buy. But to succeed the way Thad intended to succeed—that was going to be something truly amazing.

He grabbed the duffel, which was still fairly heavy, then pulled out his cell phone and dialed Rebecca's number. She answered on the first ring.

"Is it as bad as it looked in the yellow pages?" she asked, by way of a greeting.

Thad glanced back over his shoulder as he reached the door. Bed up against the wall, tarp laid out across the floor, bristling with shiny new tools.

"Actually, Rebecca, it looks fucking beautiful."

Phase Three was complete.

Houston, we have liftoff . . .

28

"I don't think anyone else is going to show up."

Thad drummed his fingers against the steering wheel as he peered up through the windshield. Even with the wipers at full blast, he couldn't see much through the swirl of fierce rain that enveloped the entire parking area. The tiny cone of orange light from the Jeep Cherokee's headlights was no match for what had now become a torrential downpour.

"Of course nobody is going to show up. It's a goddamn hurricane out there."

He turned to look at Rebecca, who was sitting in the passenger seat next to him. She was hunched forward over the dash, rubbing a hand against the condensation that was slowly spreading across the inside of the windshield. He could see that her pale hand was trembling and there was a little bit of sweat forming on her upper lip.

"It's freaking tropical, that's for sure," Sandra butted in from over Thad's shoulder. "I mean if anyone else *was* going to show up, they'd turn right around as soon as we got to the observatory. Superman couldn't see the stars through this mess."

Thad exhaled, adding to the condensation on the window, then gave Sandra a look in the rearview mirror. She was right up against the back of his seat, sitting Indian style on the flat surface they had created in the back of the Cherokee by lowering the second row of seats. She looked almost as nervous as Rebecca, though her voice didn't betray nearly as much tension. Behind her, Thad could make out the bulky form of the duffel, and the jutting metallic shape of a much larger object, which they had picked up on the way to the rendezvous point. The heavy metal thing had cost more than all the other tools combined—and the funny thing was, Thad was actually hoping they wouldn't ever need to use it. But as always, he lived for the details, and at this point, he wasn't taking any chances. Like the staging area in the cheap motel, preparation was all about planning for the things you didn't see coming. *Like an unexpected tropical storm, exploding out of nowhere, screwing up their carefully planned alibi.*

It was Rebecca who had come up with the idea of putting together the observatory run the night before they planned to pull off the heist. And Thad had easily gotten more than a dozen commitments from people—mostly co-ops and interns, but even a few older scientists who had heard him talk about his popular Utah Star Parties—who were excited by the idea of spending Saturday night gazing at the stars.

Thad and the girls had loaded up the Jeep Cherokee, bought the final piece of equipment at a specialty store Thad had found online in downtown Houston, and then headed over to the meeting point, arriving a little early so they could be there when the other cars arrived. When, by nine-thirty, it had begun to drizzle, none of them had been all that concerned. In Houston, the weather came and went so quickly the meteorologists were basically throwing darts at a map. But by nine forty-five, the drizzle had become a storm, drops the size of lizard eggs crashing against the windshield and the fiberglass top of the Jeep like they were sitting in the midst of a goddamn meteor shower.

The tension inside the Jeep seemed even more explosive. Even though the heist itself wasn't going to take place until tomorrow, they had all agreed that the Saturday-night excursion was going to make for a perfect beginning to their alibi.

The alibi was ruined, but Thad didn't feel discouraged at all; in fact, the rain splattering against the windshield, as well as the obvious tension taking hold of his two young and pretty accomplices, was giving him a palpable thrill. Even the word *alibi* excited him as he added it to the list. *Alibi, accomplices, heist.*

As the excitement reached a peak—the rain slamming down above his head, the perfumes of his two accomplices mixing with the scent of adrenaline—Thad had a sudden thought, which he immediately put into words.

"Why don't we do it now?"

The question echoed through the interior of the Jeep, for a brief moment drowning out the sound of the rain. Thad glanced over at Rebecca. She was staring at him, her hands clenched against the dashboard in front of her. God, she was beautiful. Even in the dark, broken only by the dim light from the headlights and the few blinking diodes from the dashboard—she was truly beautiful. The wash of love he felt when he looked at her filled him with strength, tripled his determination.

Sure, he had only known her a few weeks, but she was giving him an almost inhuman power, pushing him to do the impossible. But it wasn't Rebecca who broke the silence; it was Sandra.

"You're kidding, right?"

"Think about it," Thad said, his voice now a whisper. "We've got all the equipment with us. We've got the hotel room. And it's almost ten o'clock on a Saturday night. That's even better than a Sunday. Nobody's going to be there."

He was still looking at Rebecca—and then he saw a flash of brightness form behind her eyes.

"The rain is a perfect cover," she whispered. "Nobody can get a good look at the Jeep. The exterior cameras will be pretty much useless. It's kind of perfect."

Thad reached out and put his hand on top of hers. He could feel that her entire body was trembling. He started to tremble too, but not because he was afraid.

He looked into the rearview mirror, matching Sandra's gaze. Slowly, she nodded.

"Holy shit," Thad said. "We're really going to do this, aren't we?"

And then he reached for the ignition.

. . .

Rebecca was right; the rain was a perfect cover. Thad's heart was beating in tune with the oversized drops as he pulled the Jeep to a stop in front of the security kiosk at the outer gate of the JSC campus—but almost immediately, he realized that neither of the two burly guards inside was going to stick even a limb out of their warm, cozy nest. They certainly weren't going to come outside in the downpour to inspect a vehicle with a NASA sticker emblazoned on the side window.

In fact, the closest guard didn't even shine his flashlight in Thad's direction as Thad dutifully held his ID card out his half-open car window. Thad knew from experience that the guards never really looked at the pictures on the IDs, but the rain was added security. There was no way anyone inside the kiosk would be able to tell that there were three people in the Jeep; nor would they notice the large, bulky metal object in the back. And even if one of the cameras affixed to the kiosk roof, or the camera attached to the gate—which was already in the process of swinging up to let them through—managed to get a shot of the Jeep's license plate, it wouldn't make any difference. The plate wouldn't match anyone who worked at NASA, and if the authorities one day questioned the poor dude who had parked in the back corner of a strip mall the day before, they'd never connect him to Thad or his accomplices.

Once through the gate, Thad carefully pushed the Jeep to the 5 mph speed limit and began the long, crisscrossing ride to their destination. He'd always found the JSC campus speed limit annoying, but tonight, with his nerves jumping off and his stomach churning, it was almost unbearable. But the last thing he needed was a security guard pulling him over for speeding. He had a dozen stories ready in case they did run into someone—but once he'd been identified, the entire heist would be off. They were still in a place where they could turn back at any moment—they still hadn't yet crossed that invisible line that separated thought from action. But the line was getting closer by the second.

None of them uttered a word until Thad finally pulled the Jeep around to the back parking area behind Building 31, finding a spot right up in front of the mechanical bay door that was used to bring heavy lab equipment in and out. Thad had never parked this close to the building before, and it looked twice as large from where he was sitting, its rectangular frame rising up into the heavy gray rain. He turned off the ignition, shut the lights, then listened for a moment to the rain pelting off the roof and windshield. He could hear Rebecca breathing hard next to him.

"Okay," he finally said, pulling a small wrench out of the glove compartment. "Wait here."

Now that it was after ten, the air had become remarkably cool for the middle of July, even in the midst of an intense rainstorm. Thad quietly shut the Jeep's door behind him and hurried to the side of the building, pressing himself tightly against the wall so that he was partially covered by the slight overhang that extended out from the building's roof. First, he sidestepped his way to the enormous bay door that was directly behind where they had parked the Jeep. He could tell by looking at the electronic controls on the door that it was fully functional, though locked from the inside. Satisfied, he sidestepped back the way he came, then went another ten feet to a small alcove built into the wall. Within the alcove was another steel-framed door, much smaller than the cargo bay, but just as locked.

Thad passed the wrench to his left hand, then dried his right palm and fingers against his slacks. Then he reached for the electronic keypad next to the door. From memory, he entered five numbers—and smiled as the lock clicked open. So far, so good. Getting the combo for the rear door to Building 31 had been ridiculously easy. If this had taken place a year ago, he would have known the combination himself—because often, the scientists who worked in 31 used the rear entrance to get in and out of the building after hours. It was especially convenient because it was located close to the astrophotography printer room; if you needed to run off a dozen pictures of the dark side of the moon, this was where you went.

Thad had simply called one of his old acquaintances from the Monday lunch meetings a few days earlier and explained that he needed to get some pictures printed up. The man had been happy to give him the code—and had probably promptly forgotten about the call. Even if he did remember that Thad had asked for the number, it was going to be scant evidence of any wrongdoing. Dozens of people would have used that door over the past few days.

Thad bent low, placing the wrench in between the door and the frame, propping it open a few inches. Then he quickly returned to the Jeep, leaning in through the driver's-side door.

"Okay," he said, sounding much calmer than he felt. "This is going to take me about ten minutes. If you see anyone—anyone at all—just take off. If alarms go off or you hear shouts or see lights—just go. Don't wait around for me. I'll be fine."

Rebecca's eyes narrowed, and she quickly shook her head.

"No way. I'm coming with you. That was the plan."

Thad looked from her to Sandra. Yes, that had been the plan, but now that they were actually there—about to cross that line for real—he wasn't sure he wanted either of them to leave the Jeep. Even if he were caught, Rebecca and Sandra couldn't possibly get into huge trouble just for waiting outside in a Jeep. At worst, it was like a college prank gone bad, a couple of coeds cheering on an adventurous kid. Rebecca was his catalyst and his heart, but she didn't need to be his cell mate if this all went wrong.

"Thad," she said. "I want to do this with you."

He stood there, the rain pelting his shoulders and back. He was caught between the fantasy of the moment and the real-world thoughts running through his head. She *wanted* to do this with him. But he shouldn't let her. He *knew* that he shouldn't let her. But the thing was, if she did this with him, he also knew that it would bind them together in a very real way, for the rest of their lives. When two people survived something crazy—and had a secret this big to keep for the rest of their lives—it connected them in a way that nothing else could. The money was one thing: it would change their lives, it would give them the opportunity to do many wonderful things, to be scientists, to go to Africa, to be happy. But beyond the money, the experience would change them.

He already knew that he loved her, totally and intensely, but if she accompanied him on this dangerous mission, she would be just as in love with *him*. He was sure of it. They would have this forever, no matter what else happened in their lives. *They would have this.*

"Okay," he suddenly said, shocking himself—and she was already sliding across the front seat of the Jeep to join him out on the pavement. "And, Sandra—"

"Uh, I'm happy to wait in the car."

Thad smiled at her over the backseat. He turned to face his girlfriend, and squeezed her hand. Then he led her to the back of the Jeep and opened the rear door. While she stood and watched he reached inside with both hands and hefted the oblong, oddly shaped metal object and lowered it onto the pavement. Then he went for the duffel, slinging it over his shoulder. He paused a moment, ignoring the rain that was pouring down over them, going through it all in his head. Everything seemed in order.

He turned to Rebecca and gave her a confident grin.

"Time to cross that line."

A second later, they were inside Building 31.

• • •

There he was.

Up on the big screen.

Twenty feet tall in all his cinematic glory. Garbed in blue-gray overalls, his face mostly covered by a thin white surgical mask, his damp, curly hair hidden beneath a latex hospital cap, his hands gloved, even his shoes covered by white cloth booties. He was moving like a cat down one side of the never-ending hallway, his knees slightly bent to conceal his vertical motion, his bright green eyes keeping track of the revolving security cameras, making sure he was out of sight, again a ghost, a breeze, as invisible as air. His girl was right behind him, following his every move, mimicking his gait, dressed just like him, helping him drag the metal thing along the wall—its clinking, creaking wheels the only sound beyond their stifled breaths, the patter of their covered shoes against the cement floor.

A flurry of choreographed motion as they suddenly shifted to the other side of the hallway, passing from one security camera to the next, never slowing, never hesitating, moving like trained dancers across a Broadway stage.

And then, in front of them, the door with the cipher lock. Without pause, Thad reached into the duffel and retrieved a small, handheld black light that he had bought at Home Depot along with the tools. With a flick of his thumb, he engaged the light and shined it on the keypad. Rebecca gasped behind him as five of the keys lit up, bright as the moon on a cloudless night. Except, when she looked closer, she could see that the brightness was different with each key; a cascading scale of light, from the brightest number to the dimmest. Thad winked back at her, his green eyes the only part of his face visible above the surgical mask. His magic powder—the combination of fluorite, gypsum, and talcum—had worked. He had powdered all the keys—but only five numbers on the pad had been pressed within the past twenty-four hours, because the person who had pressed the pad had known the password, hadn't been guessing in the dark. And with each key he had pressed, the oils on his fingertips had absorbed a little bit of the talcum, taking a little bit less of the fluorite along with it. Thad didn't even have to guess the sequence of the five numbers—he could read it with the ease of reading five letters on a page.

One at a time, he pressed the keys in order of brightness. There was a buzz, the whir of mechanical gears—and the door clicked open.

"Okay," he whispered through the surgical mask. "Wait here."

This time, Rebecca didn't complain. Thad could see, from the look in her eyes, from the sweat that was dampening her surgical mask, that she was now terrified. Her breathing was becoming short and fast, and there was the real chance that she was going to hyperventilate if she kept it up. He leaned close to her, so close that his forehead touched hers, and stared straight into her eyes.

"This is it. This is happening. And you're going to be just fine. Stay right here; I'm going to take care of everything."

Her breathing eased, and she nodded. She was scared, but she was going to make it. She trusted him. She had reason to trust him. In her eyes, he was James Bond, the guy who could do anything, who got her to jump off cliffs and dive out of airplanes. He spoke multiple languages, swam with astronauts, and might one day walk on Mars. He was going to give her the moon.

He turned and, alone, headed through the door.

. . .

And into the Lunar Lab. Past the Plexiglas nitrogen cabinets with the attached bristle of rubber gloves, whirling, twirling, right up to the massive steel door with the immense wheel lock, spinning, spiraling, through the steel door and into the vault, careening, teetering, past the skyscraper-like steel cabinets with the aluminum drawers, staggering, tottering, through the miniature door to the safe marked *trash*, kneeling, keeling, fingers on the electronic lock, hitting the numbers one after another after another after another, and—

Nothing.

Reset. Resume. Hitting the numbers one after another after another, and—

Again, nothing.

Thad jerked back from the safe, and suddenly reality hit him like a

Saturn V rocket to the face. He wasn't in the lunar vault at all. He hadn't gone through the miniature door, or past the steel cabinets. He hadn't opened the massive, unopenable, impossible wheeled vault door.

He wasn't in the lunar vault. He was in a lab. Specifically, he was in Everett Gibson's lab, the same lab he had once visited with his wife, Sonya, so she could see a moon rock for herself. And he was standing in front of Everett Gibson's safe, staring at a combination lock that would not open.

He blinked, hard. He truly wasn't certain when the plan had changed—when, exactly, he had shifted from the mental game of breaking into the lunar vault to the much more practical, much more doable puzzle of breaking into Gibson's lab, to get to his safe. But somewhere along the way, just enough reality had broken into Thad's fantasy to push him to this place, to this crime. In his mind, standing there staring at the shoulder-high safe, which he knew contained five drawers filled with specimens that Gibson had been collecting, experimenting on, for more than thirty years—in his mind, it was morally equivalent to robbing the trash safe in the lunar vault. These were used moon rocks, stored away in this safe in a corner of a sixteen-by-twenty-foot lab, trotted out now and again for a lecture, maybe carted around to a high school or a college or a private NASA function—but they were essentially still considered NASA's trash. Gibson had had thirty years with them; it was now Thad's turn to put them to use.

And what of Gibson, what of the kindly, professorial man who had been a part of NASA's history, who had personally handled and held these moon rocks from the moment they had been brought back from the Apollo missions, who had been there when the moon landings actually took place? Well, Gibson had already lived through that experience; he'd have that glory and that moment within him for the rest of his life. Now it was Thad's turn.

Thad blinked. His mind whirled back to Rebecca, still standing out in the hallway, probably terrified, trembling, nearing her breaking point. His jaw stiffened as he tried the combination one more time. Again, nothing. He took a deep breath, then sized up the safe with his eyes. Yes, it was big, and he knew that it was heavy. Between five and six hundred pounds. It wasn't a guess, it was something he had researched, something he had hoped he wouldn't need to know—but then, again, preparation was all about the details. The things you didn't expect to need to know. Plans within plans. Thad had expected to be able to open the safe—but he had planned for the chance that he couldn't.

He spun on his heels and rushed back out through the lab, now seeing it for what it was, a somewhat cluttered place of test tubes, steel sinks, and chrome shelves. Much like the lab he had worked in for two wonderful tours. He reached the door and stuck his head out into the hallway, catching Rebecca by surprise. She jumped back, nearly tripping over her covered shoes. Thankfully, she caught herself before she toppled into the range of one of the security cameras.

"I need your help," Thad hissed. His calm demeanor was cracking, but he didn't have time to polish the rough edges. They had to move fast.

"What? Why? What's wrong?"

"Nothing's wrong. I just can't get the safe open."

Rebecca's eyes became saucers.

"You can't get the safe open? Christ, what are we going to do?"

Thad pointed past her, to the metal thing they had carted along with them from the Jeep.

"That's why we brought the dolly."

Rebecca exhaled into her surgical mask. Of course that was why they had brought the damn thing—heavy and unwieldy, but rated to six hundred pounds with a mechanical crank lift and heavy-duty straps—really, Thad had hoped they wouldn't need it. He had thought he was going to be able to open the safe right there in the lab. In fact, that was the main reason he had focused on Gibson's lab when, even in fantasy, he'd realized that the lunar vault was impregnable.

Even though Gibson had made him stay outside the lab when he had retrieved those moon rocks for Sonya a year ago, Thad had been able to see the numbers affixed to the top of the safe. He had assumed they were the safe's combination; obviously, he had been wrong. In retrospect, it was foolish to have thought that a man as smart as Gibson would have the combo right there in the lab. No doubt the numbers were actually a memory tool—maybe some sort of algorithm that helped the man calculate the combination each time he opened it. It would be easy enough to devise an algorithm that could be changed every few weeks without much effort, and you wouldn't have to memorize anything other than the process for using the algorithm—multiplication, subtraction, whatever that might be.

Given enough time, Thad knew he could probably break the sequence—but he certainly didn't have that time here and now. What he had was a dolly that was rated to six hundred pounds, an extra pair of hands—be they little, porcelain, and trembling—and a Jeep Cherokee waiting outside.

He stepped past Rebecca and grabbed the dolly, then pointed her ahead of him, into the lab. A minute later, he was back in front of the safe, Rebecca next to him. Carefully, he moved the dolly into position, then shifted so that he could put his full weight against the safe. Straining every muscle in his body, he tried to tilt it off the ground just enough to get the edge of the dolly beneath it. No dice; even with all of his weight, the damn thing wouldn't budge.

"You're going to have to help."

Rebecca quickly put her hands next to his, and together they tried again. Thad's face turned bright red, his arms and thighs becoming taut, his back crying out with the effort. Slowly, the thing creaked forward—and then it was up, just an inch, maybe two. Thad used a leg to get the dolly underneath—and then the safe crashed back down. But it had worked, the dolly was beneath the edge, and with him using both shoulders, it was only another minute before Thad had the thing where it needed to be. Quickly, he fastened the heavy-duty straps around the corners of the safe, and it was ready to go.

Grinning as he breathed hard, he tilted the dolly so that its weight was on its wheels, and slowly began dragging it back through the lab. Rebecca followed him, making sure the safe didn't tilt or twist. Every now and then, she glanced up into his eyes—and Thad could see that she, too, was grinning beneath her surgical mask.

29

"One. Two. Three. Lift!"

Teeth clenched, shoulders burning, Thad strained against the safe with all of his strength as the two girls leaned their combined weight against the handles of the dolly, and slowly the thing angled backward just enough to get it over the raised lip of the hotel doorway. A second later, they all let go at once, and the thing crashed back to the floor, rocking what felt like the entire room.

Thad exhaled, shaking the sweat out of his hair. Then he went to work on the straps. Once the safe was untied, he motioned the girls out of the way and, using a back-and-forth motion, managed to rock the safe forward so that it slipped, inch by inch, off the dolly and onto the tarplike sheets he had laid out over the carpeted floor. Once it was safely on top of the sheets, he rolled the dolly out of the way, and the three of them stood in the doorway, looking at the steel monstrosity in the middle of the room.

"Christ," Thad said.

"Yeah," Sandra responded. "That's probably not the appropriate word."

Thad smiled, putting a hand on her shoulder.

"You can play lookout. Stand outside the door, keep an eye on all the other rooms and the parking lot outside. If you see something, shout."

Sandra seemed all too happy to step outside, shutting the door behind her. They were all on edge—a mixture of excitement, but also a little fear, because now it was here with them in this room, a great monolith that seemed to suck all the oxygen out of the air. Thad could only guess how long it had sat in that corner of Gibson's lab. How many times the old man had opened that door, lovingly placing specimens inside. Well, Thad only intended to open that door once.

He crossed to the tools laid out to the side of the safe and found what he needed. A large, handheld Skil saw with a specialized blade. Looking at it, he knew immediately that the blade was too thick for what he intended to do—so it was going to take some time. Worse yet, it was also going to make some noise. A lot of noise.

"Rebecca, the TV."

"You want to watch TV?"

He shook his head.

"Just make sure it's something loud."

She blushed, understanding. She quickly rushed over to the set and turned it on, found some sitcom on one of the major stations. She turned the volume all the way up as Thad approached the safe.

Carefully, Thad placed the saw blade against the crack at the edge of the locked safe door and began drawing it back and forth—first slowly, to make sure he didn't slip, and then faster, each stroke grinding away at the blade, sending up little wisps of metal and smoke. Grinding, grinding, grinding, the sound of metal against metal a near screech in the small room, just barely covered by the inane babble from the television. He went for about fifteen minutes straight, then stopped, his arm burning, sweat running freely down his back. He signaled Rebecca, who muted the TV. Then he looked back over his shoulder toward the door.

"Anything?" he called out, in a loud whisper.

Sandra, who was standing right outside, called back.

"Nope, keep going."

And then he was back at it. The TV up, the saw a blur of motion.

Grinding, grinding, grinding. Another fifteen minutes, then pause. The TV down, the room gone silent.

"Now? Still okay?"

"Still good. This place is deserted. I don't think anyone is on this floor."

Thad grinned through his growing exhaustion, then went back at the safe. Grinding, grinding, grinding. He could see that it was working, that the saw blade was slowly thinning—soon it would fit all the way into the crack, and then he'd be able to go to work on the pins that held the lock in place. Thad knew from research on the Internet that a safe this size would have four pins. He had no idea how hard it would be to get through them—but he'd bought a good half-dozen different blades, just in case. Hopefully, they'd be done before dawn, when assuredly someone, maybe a maid or a hotel manager, might wander by. Until then, he assumed it was going to go like this for a while—fifteen-minute intervals of work, a few minutes to pause and see if anyone had overheard, then back to work.

But his assumption turned out to be incorrect; just one more break, and one minute into the back-and-forth with the saw, and there was a sudden, loud metallic pop. Thad froze, looking up at Rebecca. She quickly shut off the TV, and both of them moved close to the safe, peering into the crack.

"Holy shit. The pin—it's aluminum! It just popped like a fucking bottle cap!"

Rebecca clapped her hands. Thad quickly motioned her back to the TV and switched position, moving the saw to where he assumed the next pin would be. And again—*pop!*—just like that, he was halfway done. Within another five minutes, he'd gotten all four pins. He carefully removed the saw and placed it on the sheet, next to the other tools he wouldn't be needing. The safe hadn't been anywhere near as difficult as he had expected.

Rebecca turned off the television, and they called out to Sandra, inviting her back inside. After she'd locked the door behind her, they went to the duffel and grabbed the materials they would need. First, they all redonned their latex gloves. Then they positioned a tackle box next to the safe—oversized, metal, the kind of thing a professional fisherman might use—ready for the samples they were going to sell. Next to the tackle box they placed a small suitcase that Rebecca had brought over to the motel earlier in the day, which would be for the paperwork and anything else that might be needed to go along with the tackle box. And then next to the suitcase they unfolded the large packing box. The address was already written out on top of the box: it was a general NASA administration address, which meant it would take a few days for anyone there to process—but eventually, they would get the package and find whatever Thad and the girls sent back. Thad intended to return everything they weren't going to sell, or anything that he didn't consider trash—no matter what NASA or Gibson might have labeled it.

Finally, Rebecca retrieved a notepad and a pen. She was going to be the secretary of the event, logging and recording everything they found inside the safe, keeping everything cataloged exactly as they found it—weights, amounts, position in the safe—recording everything, just in case. They were, after all, scientists, and they were going to treat the samples with a scientist's respect.

In solemn fashion, Thad approached the safe door. He gave one last look at Rebecca, then reached for the edge and slowly pulled it open.

As he remembered, there were five drawers inside, most of them containing small containers, capsules, and Teflon-sealed bags of material. Carefully, he reached for the closest drawer, and with his gloved hands picked up the nearest container.

"Sample 167106.88. From Apollo 16. Light, clean. The lunar highlands."

He heard the sound of the pen scratching against the notepad. His mind was swirling. He was holding a vial that contained a tiny sample that had been retrieved by the astronauts of Apollo 16. It was almost unbelievable. He carefully placed the sample in the tackle box, then moved on to the next one. This was in a bag, dust with some pebble-sized pieces mixed in. It had a reddish hue.

"Vial 17422.20. Apollo 17. Brought back by astronaut Jack Schmitt—the only official geologist to ever step on the moon. The infamous orange soil. Volcanic in nature."

He placed the bag in its own compartment within the tackle box, then returned to the safe. His eyes immediately moved to one of the containers in the next drawer down, because the catalog number jumped out at him. He realized, as he read it to himself, that it was from the very first Apollo mission. It had been collected by Neil Armstrong—the first man on the moon.

Thad lifted the little container out, but he couldn't get the words to come out of his mouth. Rebecca and Sandra were both looking at him. Finally, he cleared his throat.

"This one I'm keeping."

"Thad—"

"We have more than enough to sell."

And then he had an even better idea. He was going to keep a little bit from each sample—just some dust, a pebble or two. Even after selling what they sold, he'd have the best rock collection in the world. He placed the container with Neil Armstrong's sample aside and went back to the safe.

Painstakingly, over the next hour and a half, he went through the entire top four drawers. Slowly, as he went, it began to dawn on him—and by the time he finished, he knew for sure—that in that safe, they had samples from every single moon landing in human history. Some were tiny, little more than dust. Some were bigger, but none was particularly large. Altogether, in total, the weight of the samples added up to 101.5 grams. A little less than four ounces. It was far less than Thad had thought would be inside—but it was still an incredible haul. Although the deal he had made with the Belgian was for a hundred thousand dollars' worth, if he actually wanted to calculate the full street value of what he had taken . . . well, it would have varied depending on what numbers he used—but he knew the range could be anywhere from $400,000 a gram to $5 million for the same amount. That put the value of 101.5 grams of the rocks at somewhere between $40 million . . . and half a billion.

It took Thad another thirty minutes to carefully parcel out a little bit from each sample to a separate container, which he intended to keep. It really would be the ultimate rock collection—a sample from every single moon landing there ever was, and maybe ever would be. Whatever the street value, it was actually quite priceless. Then he turned back to the safe and reached for the bottom drawer.

He recognized a few desiccators from his work back in the life sciences building, and knew from their appearance that they contained meteor fragments. Most of these, he had Rebecca put into the packing box, to send back to NASA. Toward the back of the drawer, he saw a desiccator that seemed slightly larger than the rest. Curious, he retrieved it, holding it close to his eyes as he read the label.

To his utter shock, he recognized the call letters immediately.

"ALH 84001."

He stood there, staring at the little fragment inside.

"What's that?" Sandra asked. "Another moon rock?"

Thad shook his head. Not a moon rock. It was even more valuable. It was the Mars sample—a fragment of the meteor that Everett Gibson had used to prove that there had once been life on Mars. The one that had been recovered from the ice in Antarctica in 1984.

"This one is from Mars."

"Mars? You're kidding."

He shook his head. Then he carefully placed it in the tackle box.

"Why are you putting it there? Are we gonna sell it, too?"

"Maybe," he responded, though he didn't think he ever could. But for some reason, he wanted to take it with them as well. God knew how much it was worth to a collector like the Belgian; truthfully, Thad didn't know if any amount of money would persuade him to let that one go. The idea that he now owned a piece of Mars was hard to get past.

"Okay, now for the paperwork."

Beneath the bottom drawer, Thad found the curatorial forms—the actual NASA log of all the samples they now had in the tackle box. It was the best receipt—and written proof of the samples' authenticity—that they could ask for. Thad carefully placed the forms into the suitcase, along with everything else that looked important still within the safe—a few loose papers, a vial or two—and was about to go about the process of reorganizing the tackle box by mission, in sequential order, when he noticed that Rebecca was still focused on the bottom drawer of the safe.

"Thad, what about that? That dust?"

Thad peered into the safe and saw what she was pointing toward. In one corner of the bottom drawer, there was a tiny bit of reddish-white powder. He realized that sometime during the process of moving the safe on and off the dolly, one of the sample bags must have leaked a little bit. It was really just a tiny amount—less than a gram, a very fine layer in just one little corner of the safe—but it was still from the moon. Thad stood there, thinking about it for a few more seconds—and then he did the only thing that came to mind.

He took his finger and slid it through the dust, then placed it into his mouth. Swallowing, he then grinned back at Rebecca.

"Now I'll have a bit of the moon inside me."

Without waiting for her reaction, he sealed the tackle box, closed the suitcase, and began cleaning up the rest of their staging area. He loaded the tools into the now-empty safe and then retrieved the dolly from where they had left it, near the bureau with the TV. The exhaustion was really starting to hit him, but he knew they still had a lot of work to do before the night was done.

As Sandra helped him work the safe back onto the dolly, Rebecca carefully folded up the sheets, then gathered the tackle box, suitcase, and the package addressed to NASA, and followed them toward the door. Thad and Sandra fought to work the still immensely heavy beast over the door frame. Meanwhile, Rebecca couldn't help but ask the question that was on both girls' minds.

"So how did it taste?"

Thad grunted as the safe lurched over the door frame, then inched its way outside.

"Salty, actually."

He realized as he went that he was probably the only person on Earth who could say that with authority.

. . .

"I think I'll have the Grand Slam Breakfast. In fact, we'll all have the Grand Slam Breakfast. Grand Slam Breakfasts all around!"

Thad knew he sounded ridiculous, but he couldn't help himself. Besides, if you couldn't sound ridiculous in a deserted Denny's situated on a lonely stretch of highway somewhere in the middle of bum-fuck Texas, then where could a guy, his girlfriend, and his confidante go to let off steam? And besides, it was really late—he wasn't even sure what time, just that it was really freaking late—and he was beyond tired, so punch-drunk from living off that adrenaline high for so long that for the first time that he could remember, he had limited control of his faculties.

The girls didn't seem much better off. Rebecca, for her part, had turned twice as bubbly as usual, and she was downing diet Coke after diet Coke as she counted—out loud—the rare headlights that flashed by on the highway outside the large picture window behind them. Sandra's eyes were half shut, and she was slouched over at the banquette-style table, halfway between awake and asleep. Every now and then, Rebecca kicked her under the table just to make sure she was still conscious.

Hell, it had been a long night. But as far as Thad could see, they had done everything perfectly, and had taken every precaution. After leaving the hotel room, they had first taken to disposing of the safe. Driving to a small town called Alvin, Texas—a good forty minutes from the outskirts of Houston—they had wandered around until they'd found the perfect Dumpster, in a deserted alley next to an oversized car park. They'd chosen a separate Dumpster, another town away, for the rest of their trash, including the safe door, which they'd removed in the back of the car—since it had only been hanging on by one of the twisted pins by that point—just for good measure. Then they drove back to the old car that Thad had taken the license plate from, returning it to its rightful owner.

And then they had done the only thing they could think of to cool down: they had gone for breakfast. Denny's was the only restaurant open that late, so Grand Slams it would have to be. And as tired as they all were, they knew that things would be even rougher over the next twenty-four hours as they completed their plan.

As the final part of their alibi, Thad had arranged to shepherd a group of co-ops to a famous Texan water park the next day. It was going to take a true force of will to make it through the excursion—not just because of their exhaustion, but because of the secret they now shared, the secret that was so fantastic, so unbelievable, it was going to be a real feat to keep inside.

And after tomorrow, well, there would be an entire five days of business as usual, all of them back at their routines. Thad had made the final arrangements with Lynn Briley the day before via e-mail; he would be meeting the Belgian's sister-in-law in Florida the following Saturday, five days away. Instead of Tampa, they had agreed on Orlando, because Briley had suggested it. Thad had never been to Orlando, but he didn't think he'd have much trouble finding the meeting place—a restaurant on International Drive called Italliani. The drive down to Florida would be more of a strain; but at least then, he and the girls would be able to talk about what they had just done, as much as they wanted.

Although, to be truthful, Thad liked the fact that for the moment, in public, it had to remain unspoken, this incredible secret. Leaning back against the banquette, he smiled at Rebecca, who twirled her straw at him, smiling back. He knew the secret of the moon rocks would bond them forever. Long after they sold them, and mailed back the excess to NASA, they would have the experience they had just gone through—something they would cherish, remember, and dwell on forever. He loved her with all of his heart, and he felt certain she loved him back.

In less than a week, they would have enough money to go away together, maybe to really start a life with each other. In his mind, Sonya was the past; Utah was the past; his family, Mormonism, even Everett Gibson—the past.

Rebecca was his future. A hundred thousand dollars in a briefcase was his future. And a little fragment of the planet Mars.

All of this was his future—*and the future was fucking beautiful.*

30

Eyes closed, head down, Thad braced his hands against the sides of the glowing white cubicle and let the superheated jets of water pummel his naked shoulders, neck, and back as the steam from the nozzles embedded in the floor beneath his feet billowed upward in glistening, amorphous clouds, filling his nostrils, mouth, and lungs. More jets on either side spat powerful streams of even hotter water at his sides and chest, the angry rivulets tearing at his skin like white-hot needles, carving a grimace onto his lips and a wince into the edges of his eyes. But still, he didn't move, letting the computer that controlled the space-age shower's temperature and water pressure continue along the brutal preprogrammed cycle, hotter and hotter still—until there was a near scream working upward through his throat.

And just then, thankfully—when he knew he wouldn't be able to take it anymore—the water suddenly shot off, the steam whirling upward into the vented grates that lined the brightly lit ceiling panels. Thad stood there, naked and dripping steaming beads of nearly gaseous H_2O, gasping for the cooler air that now made its way into the shower cubicle. Christ, that had been intense—but it was exactly what he had needed. Not only to work the knots out of his strained muscles, but also to clear the nearly constant state of tension from his brain. Even though it had been two days since the heist, his entire being still felt clenched, like a steel spring compressed so tight and flat that he was liable to explode. Luckily, it was only going to be another few days—the exchange with the Belgian rock hound's sister-in-law had been confirmed, and Friday afternoon he would begin the long drive down to Orlando, Florida.

Which meant he only needed to blunder his way through his regular NASA routine for a little while longer. It was eleven A.M. on a Tuesday, and he was exactly where he was supposed to be, the shower room of the NBL, wasting time as he waited the necessary hour before the doctor could check him out for his lunch break. Sure, maybe he had dawdled a bit longer than usual in the Jetson-family shower, but he was sure nobody was going to notice.

Hell, it had been almost three days, and nobody at NASA had yet noticed a six-hundred-pound safe missing from the lab of one of Building 31's premier scientists. He doubted any newspaper reporters would be sent out because a co-op had been getting a little too friendly with a few hundred computer-controlled shower nozzles.

He finally opened his eyes, shaking the water from his hair. As he stepped out of the cubicle—and watched with his regular sense of awe as the folded hot towel slid out of the wall in front of him—he reminded himself that whatever tension he was feeling, he knew that the girls were probably in much worse shape. He had been living with the heist as a mental image for more than a year; Rebecca and Sandra were probably neurotic messes by now.

Thad had spent as much time as possible calming them down via telephone and through protracted lunches; he had gotten them both to the point where they were no longer staring at the door, expecting armed police to come barreling in at any moment. But he was still extremely glad he had planned the trip to Florida for such a close date. He doubted either of them would've been able to handle another week like this.

Wrapping a towel around his waist, he crossed to his locker and was about to reach for his clothes when there was a rush of air behind him, followed by the sound of skidding footsteps.

"Did you hear the news?"

Brian was halfway into the doorway of the locker area, an excited look on his face. He was still wearing his wet suit, slung down around his waist, his bone communicator hanging over one shoulder. He'd obviously just come from the NBL deck, though they had both gotten out of the water at the same time almost an hour ago.

"I haven't heard anything. I've been in the shower—"

"You look like you just stepped out of a pizza oven. You never heard the story about the frog and the pot of water—you know, he doesn't realize he's boiling until it's too late?"

"Is that the news? They're boiling frogs over in the NBL?"

Brian shook his head, stepping all the way into the locker room.

"You're not gonna believe it. Someone stole six hundred pounds of moon rocks."

Thad's stomach dropped. He was glad he was sitting down. He was also glad that his skin was still bright red from the hot water, because he was certain that otherwise, his cheeks would have turned as pale as Rebecca's.

He'd prepared for this moment—sooner or later someone was going to notice the missing safe—but it was still terrifying to hear it out loud. He didn't quite know what to say in response, but it didn't matter, because Brian was still going at a million miles per hour.

"I wish I was the smart motherfucker who thought of that. Six hundred pounds of moon rocks? You have any idea how much that's worth?"

Thad pulled a corner of the oversized towel over his head, as if he were drying his hair. Beneath the towel he was grinning. It was amazing to hear a comment like that from his friend, because it was something he would never have expected Brian to say. Brian was as straitlaced as they came. He almost wanted to tell Brian the truth. But there were already enough people involved in the situation; he wasn't going to risk adding one more.

"I'm sure it's a lot," he finally answered, his voice muffled by the towel.

"We were calculating it out over on the NBL deck. Six hundred pounds of rock—it's like over one trillion dollars."

Thad wanted to correct Brian; of course, it wasn't six hundred pounds of moon rock. It was just a six-hundred-pound safe. And it wasn't a trillion dollars, but it sure as hell was worth a lot. Then more of what Brian had just said made its way into his jumbled thoughts.

Obviously, it wasn't just Brian who knew about the missing safe. And if it had made its way to the NBL, which was ten minutes away from the campus . . .

"Everyone is talking about it," Brian continued, putting words to Thad's thoughts. "They stole the safe from Everett Gibson's lab. Gibson is still out of town, so nobody yet knows for sure what else was in the thing, but the rumors are flying. Nothing like this has ever happened at NASA before."

Thad was about to say something in response, maybe ask some more questions to get some more information, when he realized that his cell phone was ringing from within his locker. Keeping his heart rate under control, he nonchalantly retrieved the phone from the pocket of his jeans and checked the number. *Rebecca.*

As Brian continued to ramble on about the enormity of what had just happened, Thad put the phone to his ear, cupping it slightly with his hand to make sure that Brian couldn't hear the voice on the other end. Before he could get even a word out, Rebecca was half shouting at him, her voice high-pitched and obviously filled with real fear.

"Everyone knows that the safe is gone. I've gotten, like, dozens of e-mails, from people all over the campus."

She sounded frantic. Her voice was cracking at the edges, and Thad could tell she'd been crying. He wanted to tell her to remain calm, that of course people were going to find out about the missing safe, that there was no way they were going to link it to the three of them—

but with Brian standing right in front of him, he had to be extremely careful.

"Yeah, I just heard from Brian. Pretty crazy. Nobody has any idea who could've pulled something like this off. Hell, they'll probably never catch whoever did this."

"Thad, I don't want the stuff in my apartment. We need to move it now."

Thad realized that the reason Rebecca was so terrified wasn't just that the rumors were flying—but that the moon rocks were in her apartment. They had left them there because that's where they had spent the past few nights.

"Okay, yeah, that's something we can deal with—"

"Sandra says she has the perfect place. A buddy who's gone to Europe for the rest of the summer gave her a key to a storage unit. Get over here as soon as you can, and we'll move the stuff there."

With that, Rebecca hung up. Thad placed the phone back in his pants and started to get dressed. Brian sat down on the bench next to him, still alive with thoughts of the stolen safe.

"Goddamn, man, you steal something that valuable, you never go to jail. Because you can just buy off anyone who wants to turn you in."

Thad smiled at the joke, but inside, the spring had just tightened another few notches. No space-age shower would be hot enough to help him now.

31

The rumors were still swirling three days later as Thad fought his way through a perfectly typical day at the JSC; the morning spent at the NBL, scuba till noon, shower, lunch, scuba till the late afternoon, shower, then get the doctor's okay to check out—and finally, he was outside, waving good-bye to Brian for the weekend, the sun flashing against his face, curls of his still-damp hair bouncing across his forehead. He rushed back to Rebecca's car, which he'd borrowed to replace the Toyota—which had been acting funky—started up the engine, and headed across town to pick up Rebecca for the fourteen-hour drive to Orlando. Thad had calculated it all in his spare time at the NBL, leaning low over the computer as he used a variety of mapping Web sites to find the optimum route. About 950 miles, most of it highways, maybe the longest car trip he had ever taken without stopping—but there wouldn't be time to stop, and besides, and more important, he didn't have enough money to plan out any stops. Crazy, that there were many millions of dollars' worth of moon rocks in his trunk, but he couldn't afford a hotel room, or even a real, upscale restaurant. He'd just have to rely on what was in the trunk—and the memory of how it got there—to impress his girlfriend along the way.

He was smiling as he made the fifteen-minute trip over to Rebecca's apartment. The thought of fourteen hours alone with her in the car was thrilling; the fact that Sandra was unexpectedly going to be staying behind in Houston seemed like an incredible stroke of luck. She had wanted to come along, but her scuba certification test had happened to fall that particular weekend—so there really wasn't any choice. She needed the scuba cert to continue her quest to become a NASA employee, and despite what they had just done, she still intended to go about her business as usual. As Thad had spent the past week convincing both girls that business as usual had to be their primary demeanor until the heat and rumors had blown over, he hadn't been able to argue with her. And truthfully, he wouldn't have anyway.

Rebecca was already outside her apartment, sitting on the front steps, when he pulled to a stop by the curb. She grabbed a backpack from next to her feet and slung it over her shoulder, then headed toward him. She looked so fresh and happy, like it really was the first day of the rest of their lives. She was wearing shorts, like him, and a T-shirt with some rock band's logo emblazoned across the chest. She looked even younger than she was, some sort of gorgeous, sable-haired sprite infusing life into everything she got near. The fear and tension that had been visible in her for the entire past week seemed to be gone, now that they were on their way to Florida, and as she yanked open the passenger-side door, tossed her backpack into the backseat, and slid inside, Thad wanted to grab her in both hands and tear off her clothes. Of course, there would be plenty of time for that later. Instead, it was she who leaned toward him, planting a fierce kiss on his lips, running a hand down his chest to his pants, giving him a foreshadowing little squeeze. Then she got busy putting on her seat belt with one hand while unfolding a printed-out map from her pocket with the other. Her face was all business, and Thad realized he couldn't stop smiling, watching her.

It was exactly what he had thought—the experience they'd had, the secret they'd shared, had accelerated their relationship, bonding them together. It was like they'd been in love for years, even though it had actually been only weeks.

"Too bad about Sandra," she said as he started up the car. "But it's kind of nice, isn't it? Just me and you? Through to the end?"

Obviously, she had been thinking along the same lines as he—and that excited him even more. But though the drive down to Florida would just be the two of them, it wasn't going to be that way through to the end. Thad decided, for the moment, to leave her in the dark about the accomplice who would be taking Sandra's place when they reached Orlando; Thad wasn't exactly thrilled about the substitution himself, and he wasn't even sure the dude would show up.

Thad had sent Gordon an e-mail shortly after the heist, more as a courtesy than anything else. Although Gordon had been instrumental in finding a buyer, he was still little more than a drugged-out acquaintance, a link to an underworld that Thad could only picture in his fantasies. Thad had assured Gordon, time and again, that he would get his 10 percent of whatever they got in the deal—$10,000 for finding an e-mail, which seemed like a pretty good bargain to Thad. But having Gordon actually there, with him—that had never really been part of the plan.

In fact, he was still pretty sure that Gordon had no real idea what Thad had actually been up to. After all, this was the same kid who scoffed at the idea that man had made it to the moon—and it was doubtful, even now, that he knew Thad worked at NASA. Having Gordon in the same room as Rebecca seemed unnecessary and unpleasant. But Thad had sent the e-mail anyway, expecting little more than a congratulations.

When he hadn't immediately heard back, he'd sent a follow-up e-mail, fully assuming that Gordon was leaving the exchange to Thad:

> I haven't heard from you, so I'm going to assume that you are not going to Florida. So just to catch you up. The Items have been quired. There are approximately 100 samples with an average mass of .8 grams . . .

This time, Thad did get a response, but it was so strange—even for Gordon—that he had assumed it was just getting bounced back to him, the message some sort of recorded amusement that would make sense only in Gordon's fractalized mind:

Vertical vacationing. Look high in the sky if you are to find me. Off fishing. Wild horses couldn't drag me away.

But a phone call after that had cleared things up—much to Thad's dismay. Gordon was indeed planning to come to Florida, and the previous e-mail hadn't been some sort of bounce-back message. It had just been Gordon being Gordon. "Vertical vacationing" had meant he was going to go flying. "High in the sky" referred to the airplane. "Off fishing" and "wild horses" meant he would be doing stuff that he enjoyed.

Gordon was going to follow up the phone call with an e-mail with his flight details, so there was no avoiding it, Gordon was going to be part of the story. Thad wasn't sure why the kid wanted to be there physically; it was definitely more dangerous for him to get involved to that degree. All he'd really done so far was send out a bunch of e-mails. But Gordon was insistent. Deciding not to argue the situation, Thad had instead found a way to make use of his buddy once again. Gordon had gotten his mother to pay for the plane ticket to Florida by telling her he was on his way to an interview for graduate school. Thad had him also tell his mom that he needed a hotel room in Orlando. There was a Sheraton pretty close to the restaurant on International Drive, which seemed perfect.

Thad would have rather it had just been himself and Rebecca all the way, but Gordon would be the third accomplice—and there would also be a new staging area where they could make their final arrangements. Thad planned to leave the goods in the hotel when he went to the restaurant, for security's sake.

First, there would be the fourteen-hour drive, with plenty of opportunities to let Rebecca know about Gordon—and also plenty of time for him to just enjoy being with her. Two people fully in love, sharing a secret, on their way to a historic event. In four short weeks, they had lived what most couples wouldn't experience in a lifetime.

. . .

"I have to admit, it feels a little wrong."

Rebecca snuggled into Thad's chest as he wrapped his arms around her waist, pulling her tightly into the spoon of his body. His right leg rested against her thigh as they both stared out through the rear window of the car, his gaze in sync with her own. Although they were parked at the very back of the empty parking lot, he had no trouble making out the two-story Baptist church, especially the cross, which rose out of one of two humble steeples, casting a shadow—backlit by the moon—that ended just a few feet from where they were lying across the backseat of the car.

"Okay, I get your point. But you know, these places aren't just about worship. They're also supposed to be about forgiveness, peace, love, asylum. And really, we don't have that much of a choice. It's either here, or the parking lot of a Waffle House."

Rebecca playfully smacked her palm against his face. Then she turned so that she was looking past the church, to the moon behind the cross.

"Asylum I get, but we're not really asking for forgiveness, are we?"

"There's nobody to ask forgiveness from. NASA? We took a quarter pound of moon rock. They've got eight hundred and fifty more pounds locked away in the vault. The Apollo astronauts? Heck, the theft will probably bring more attention to what they did, and what NASA hopes to do next, than those rocks ever would locked away in a garbage safe. Science? With the money we're going to get, we'll be able to travel the world, build our own lab, become better scientists, maybe even astronauts ourselves."

Thad was pretty sure he sounded naive, and a little foolish, but he felt that the words were sincere. An outsider might call it all attempts at rationalization, trying to explain himself in ways that went beyond the crassness of money or the cliché of love, but in that moment, camped out in the parking lot of a Baptist church because they didn't have a hundred dollars between them to rent a hotel room—even though there were millions of dollars in moon rocks in the trunk—it was probably okay to sound a little foolish.

"And what about Everett Gibson? I mean, we stole his safe. There were papers inside that probably meant something to him. He'd been working on those samples for thirty years. He was kind of a mentor of yours—"

"Dr. Gibson is going to be just fine. I'm sure he's got duplicates of anything that ended up in that safe. And if not . . . well, science is a living, moving thing. It's not something to be shoved away in a corner. Gibson was a part of the greatest scientific adventure in human history. He had his moment, he lived his moment, and now we're taking that baton. Plus, we boxed up everything else to mail back to NASA, so he will get back what we don't sell."

Rebecca went silent in his arms. Maybe she was contemplating what he was saying, or maybe she was just looking at the moon. He was pretty sure that she wasn't asking these questions because she felt guilty—just nervous. They were really close now; when the sun came up they were going to meet Gordon in the lobby of the Sheraton Hotel.

During the long drive, he and Rebecca had talked about what would happen if somehow things went wrong, and Thad had been clear about one thing. No matter what happened, Rebecca would not get into any trouble for any of this. If Thad got caught, she would tell the authorities that she had known nothing about what was in the trunk of her car. She would play dumb, and Thad would back her up.

In return, she would be there to bail him out of jail—and even if he got in real trouble, went to trial, she would stay safe. He knew they had stolen something valuable, but they hadn't hurt anyone; it was really just a big college prank. NASA wouldn't see it that way, but Thad wasn't going to be spending the rest of his life in jail because of four ounces of moon rock.

There was no reason for Rebecca to feel guilty or afraid. She was his catalyst, he loved her—but it was his mental game that had turned real, his plan that they had followed. And he was ready to see it to its conclusion.

“We don’t need to say anything,” he whispered in her ear. “Or do anything. We’ll just lie here and watch the moon, until the sun takes its place.”

And that’s exactly what they did.

32

The Sheraton Hotel, Orlando, Florida, July 20, 2002

Wild,
Wild horses,
Couldn't keep me away . . .

The hotel lobby pitched hard to the left, then dipped forward, the carpet seeming to ripple up beneath Gordon's boots, like ocean waves licking across a sandy beach, and he tried to stand perfectly still, eyes blinking rapidly as he fought the urge to topple over. Because toppling over in a Sheraton lobby at four in the afternoon just wasn't done—no, it fucking wasn't. That was the kind of thing that drew attention to yourself—yes, it fucking was—and the last thing that Gordon needed at that particular moment was attention.

The lobby slowly began to stabilize, and soon Gordon felt okay enough to take a tentative step toward the pair of overstuffed couches that overlooked the arched doorway leading out onto International Drive. He had to admit, as he inched forward over the still-oscillating carpet, that it was a pretty darn nice lobby, for a Sheraton. He'd only been in Orlando for a couple of hours, but he was really quite impressed with the place. And a hundred degrees with a hundred percent humidity didn't feel all that bad—that is, when you had enough marijuana coursing through your system to put a bull elephant in a smiling mood.

And there, he'd made it to the couch; now it was just a matter of getting his knees bent, his ass into those friendly-looking cushions, his boots up on the pretty glass coffee table. Nothing to see here, nobody special, just a guy in a hotel lobby waiting for a couple of friends. Okay, he was a bit stoned and he'd had a couple of drinks at the airport, and he was certainly planning to have a couple more drinks and some more smokes before the day was out, but that didn't make him all that different from anybody else . . . hell, everyone was a little bit high on something, everyone had his poison.

Like Thad, or Orb, or whatever the hell Gordon was supposed to be calling him. Thad was just as high as he was, even if the kid hadn't touched pot or booze in his life. He was high on that chick, and he was high on the idea of the money they were going to make—hell, he was high on the information Gordon had already given him. The Belgian rock man and his sister-in-law, the lady who was going to be meeting them, just two hours from now. Yeah, Thad was high on all that; he was so high that he was right up there near the chandelier that hung from the lobby ceiling, so wonderfully crystal and glowing and warm, looking down on Gordon, little old nothing of a Gordon. And Gordon was down there way below, in that dark, dark place, in a well of . . . well, sadness.

Still thinking about his wife and child and sister, poor dead sister, and the world, yeah, the fucking world coming to an end. Any minute, any day, and it couldn't happen fast enough for his liking. Armageddon. Damn, but it was taking too long, like Thad and the girl, taking forever to get to the goddamn lobby. Gordon knew he couldn't wait much longer, because his high was starting to wane, and he needed another hit of something, anything, to keep it going. Because his plan was getting cloudy, and he was beginning to see that it wasn't really a very good plan anyway. Come down to Florida, be a part of something big and fun and cool, feel like a person again, alive, and maybe get the opportunity to keep on going like that. Maybe make friends with the lady and go off to meet her brother-in-law in Amsterdam, backpack across Europe with the 10 Gs he'd make from selling that moon rock, use the 10 Gs in a very responsible and intelligent manner, get some more pot, some heroin, enough heroin to OD in some Dutch youth hostel, jacked up with a needle in his arm and a rubber rope around his biceps, vein popping up, and they'd find him like that and tell his mom that he went out happy, and he'd be where he was supposed to be. Wild fucking horses . . .

And then there they were, coming through the front entrance of the Sheraton. Thad, in shorts and a collared shirt, carrying a fishing-tackle box in one hand and a suitcase in the other. And next to him, the chick, the *chica*, the Eve to his Adam. Yeah, she was pretty and had jet-black hair and was all-American and all that. And she had that greedy little look in her eyes that he now suddenly saw in Thad's as well, that greedy little cartoon look, dollar signs springing out so high they could touch the chandelier.

Four o'clock, right on schedule. Gordon waited until they were just a few feet away before he sprang to his feet. For a brief moment, he tipped left, then right, but his boots were pretty well planted in the ocean of a carpet, and not even wild, wild horses would drag him away . . .

He pulled the room key out of his pocket, showing them the number for no apparent reason other than that it seemed relevant; 905, lucky 905. And then Thad led the way, because he was a natural fucking leader, for sure, for certain, and Gordon still had plenty of jambo juicing through him, enough to make him the good little follower he needed to be. He got into step behind the girl, focused on her dark hair, because it was pretty and shiny and it would have looked interesting affixed to the rear of one of those wild, wild horses . . .

And then, somehow, they were upstairs on the ninth floor and moving through a hallway and through a door and the door was locked behind them, and Thad was placing the tackle box on a coffee table in the middle of the room, and then he was fiddling with the locks, and then it was open, and then—

Well, fucky me.

Gordon approached the table. Thad moved aside so he could look into the tackle box, and what he saw made the sober part of his mind freeze up.

The box was full of single-ounce vials and bags containing what appeared to be, from Gordon's Internet research, lunar samples. As he stood there, staring, Thad explained that they were samples from every Apollo space mission from 1969 to 1974. That although Gordon was pretty sure man had never been to the moon and it was all a goddamn hoax, he was looking at moon rocks that had been brought back to Earth by men in space suits. And then Thad pointed to another thing in the box, another sample that wasn't a moon rock, that was, Thad explained, a piece of the famed Mars rock found in Antarctica, the one that had proved that there might once have been life on the red, red planet.

"Yeah," Thad happily exclaimed. "That one alone might be worth five million to the right buyer!"

Gordon looked at him, then at the girl who was standing a few feet away, grinning some perfect-looking little teeth, and then back at the tackle box. His head was spinning, and not just the orbit of pot and booze, but the cycles and rotations of a confusion much more serious. Because these little bags also had numbers and letters on them, and the numbers and letters looked like the kind of thing that meant they were from NASA, the space agency, the *government* space agency.

"Wow, really" was all Gordon could manage out loud, but internally he was imploding. He now knew, for a fact, that there were no South American royalty trying to make ends meet, and if he was going to be honest with himself, maybe he had always known this. But at most, he had figured Thad was going to be getting a big fat moon rock from some museum, maybe the University of Utah, maybe somewhere else. And yeah, that would be illegal, sure, Gordon was helping to sell contraband—but this?

"Yeah, wow," Gordon repeated. "You guys are really serious. I thought it was going to be a sample or two—wow."

And then Thad was suddenly talking, a mile a minute, telling them both what was going to happen next. Thad was saying that first he was going to go to Wal-Mart and get some more gloves so that the buyer would be able to touch the samples if she wanted. And then he was saying that he would go to the restaurant by himself, that Gordon and the chick would wait here or go to a movie or take a swim, whatever, wait it out—and then when he brought the buyer back to the hotel, made the deal, they could rejoin him and divide up the cash. And then the chick was suddenly arguing with him, which seemed to come as a surprise to Thad; she was saying she wanted to go along to the restaurant, that she couldn't go to a fucking movie, that hell, they could make a movie out of her life—except she said it backward because she was so full of adrenaline and energy and yeah, fucking greed, she said hell, they could make a life out of my movie, and maybe she meant it that way. Maybe it sounded better that way. And Gordon was listening to it all, but he wasn't listening; he was staring at the moon rocks and knowing, just knowing that this was going to end badly, that they were going to get caught. But Thad and the chick just kept on going, and then their argument ended and Thad was agreeing and the new plan emerged: Thad would go in first and then Gordon and the chick would come in twenty minutes later like they were a couple, hand in hand, Mr. and Mrs. Americana, pretty little thing and her hubby, and they would eventually all do the deal together. And then Thad and the girl weren't talking anymore, they were just looking at Gordon, waiting for him to say something. And he was still staring at the tackle box and the moon rocks.

And it hit him, right there and then, that okay, this wasn't the way out of the well, this was the way even deeper into the well, but it was okay, it was fine, it was too late to back out now.

"So, yeah," he said, finally. "I'm going to go get something to eat."

And just like that, he was heading toward the door. Thad and the chick looked at each other and then Thad was talking low to him.

"You okay, man?"

"Sure, fine. Just going to get something to eat, and then I'll be back. Gonna get a little pizza."

And just like that, he was out the door. Moving down the hallway, using the walls for balance because the floor wouldn't stay still. Heading for the elevator, which he knew he should have taken all the way to the roof, like a rocket ship, baby, all the way out the top of the building and up into the sky. He should have taken that elevator wherever it would go, away from here, never look back. He should have simply disappeared.

But he also knew that what he was going to do, in fact, was get a pizza, maybe get a little high, and head right back to the hotel to see this through.

Wild horses couldn't drag him away . . .

33

"Gordon's going to be back any minute." Rebecca's voice drifted out through the open bathroom door, barely audible over the sound of the shower. "This could get really awkward."

Thad grinned as he yanked back the heavy blanket of the hotel bed and tested the overly springy mattress. He was stark naked, still dripping wet from the shower himself; it hadn't been the first time he'd taken a shower with Rebecca, but it had certainly been the most exciting, the two of them taking turns under the quirky, chrome-plated faucet, their bodies pressed together as his hands roamed over her skin, his fingertips gliding across her flat stomach, around her thin waist to the small arch of her back, to the gentle hills of her perfect little ass—he almost took her right there, under the low-pressure stream of water, soapy, glistening, slipping around in their bare feet against the shiny white tub—but then he had a better idea and, without a word, had dived past the plastic shower curtain and out into the hotel room.

"You obviously haven't spent a lot of time with stoners. Getting a pizza to them is kind of like a religious affair. If Gordon makes it back in time for the exchange, I'll be shocked."

Thad wished the words were true, even as he said them. They still had about an hour before they had to meet the buyer at the restaurant, and he was pretty certain from the way Gordon had reacted to seeing what was in the tackle box that the dude was going to see this through to the end.

Gordon had been pretty damn shocked at the sight of the little containers of moon rock, but even though he'd seemed shaken by the reality of the situation, he also appeared to understand the historical nature of what Thad had done. This was a party Gordon wasn't going to miss. The scary thing was, he looked stoned out of his mind, and would probably come back from his pizza mission even more so; Thad could only hope that Gordon would keep it together long enough not to screw up the deal.

Whether Gordon was returning or not, Thad knew that he and Rebecca had some time alone. A quick nap after the fourteen hours spent in the car—and the five hours in the Baptist church parking lot—would have been the most sensible thing, but Thad had come up with a much better idea.

He quickly crossed to the bureau, where the tackle box was sitting between the hotel television and the suitcase they had brought with them from Houston. Thad went straight for the tackle box, opening the clasp with almost loving care. He surveyed the carefully lined-up bags and vials containing the lunar samples. Then he reached for the bag with the markings that indicated it was from Apollo 11, Neil Armstrong's first walk on the moon.

Slowly, like he was walking down the aisle of a church, he crossed back to the bed. With one hand he lifted up the mattress cover, and then he carefully placed the bag containing the lunar sample underneath. He replaced the mattress cover and the sheet, then went back and closed the tackle box.

He was just stepping away from the bureau as Rebecca came out of the bathroom, wrapped only in a towel that was way too small for even her diminutive frame. Her porcelain skin was glistening where it was visible above the top of the white cotton material, beaded drops of water resting in the small crevice between the tops of her breasts, like pearls escaped from a necklace that Thad might soon be able to afford. Her legs, tight and muscled, were naked to the very peaks of her thighs—and even a little higher. Her hair was soaking wet, a few errant strands plastered down against the sides of her neck, jet-black strands beckoning down toward her bare shoulders and beyond, toward her perfectly sculpted back.

She was waiting for his cue. If they had had more time, he would have been content to just stand there, looking at her. But in less than an hour, they were going to be meeting the sister-in-law of a Belgian rock collector to make a deal. So instead, he headed for the bed.

If Rebecca noticed the small, fist-sized lump beneath the mattress cover, she didn't say anything. Maybe she was simply too busy, her lips against his as her hands moved low, first touching herself and then him, teasing, and then guiding. Thad's entire body surged, every nerve ending firing off as he rolled on top of her, his knees parting her legs, his hands reaching for her wrists. As the moment approached, he looked right into her eyes.

For the briefest of seconds he saw himself, hovering over her, fantasy and reality superimposed—but now the fantasy *was* real, the *moment* was real. They were making love in a Sheraton Hotel in Orlando, Florida, separated by a thin strip of material from a piece of the moon.

It was a first for humankind. Exactly thirty years earlier, to the day, Neil Armstrong had taken the first step—but right then, right now—Thad Roberts was the first man to have sex on the moon.

34

Thad did his best to conjure up the theme songs to either *Mission: Impossible* or James Bond as he strolled along the edge of the highway, but the notes just wouldn't come, his mind simply couldn't focus past the image of the restaurant parking lot—which he could already make out over a low hedge embankment a dozen yards ahead. The never-ending stream of cars whizzing by, some so close he could feel the hot wind of their exhaust against the back of his neck, didn't help; the roar of engines mixing with the metronomic beat of his own sneakers against Florida-hot asphalt was the only score he was going to get as he made his approach.

Getting dropped off two blocks from Italliani was about the only part of the newly reformatted plan that he actually liked. When Gordon had returned to the hotel room, just ten minutes ago, Thad had practically begged the two of them to let him handle the deal on his own; there was no need for them to be in the restaurant, and it seemed like such a stupid risk. His plan to protect Rebecca, no matter what, would be seriously hampered if she were caught with him, red-handed. And then there was the added loose-cannon factor, Gordon. The guy had seemed even more high when he returned from his pizza expedition, and there was no telling what he would do in the restaurant.

As it was, Thad had practically ordered them to wait at least ten minutes before entering the place, and they had agreed to play the part of a couple who'd just happened to wander into the restaurant—without any connection to Thad. If things went well, and Thad felt comfortable with the Belgian's sister-in-law and her husband, he'd call them over and together they could all return to the hotel to show the buyer the moon rocks.

It wasn't ideal, but it would have to do. Steeling himself—without the help of a really good theme song—Thad skirted past the low hedge and across the crowded parking lot.

As he stepped through the front entrance of the restaurant, he did his best to take in all the details at once—the kitschy, Italian decor, the red-brown curtains that obscured the glass picture windows, the low booths that lined three walls of the rectangular space, the waiters and waitresses wearing black and white, the hostess stand where a young woman stood talking to a pair of middle-aged-looking customers. It was the kind of restaurant that could have appeared in any town in America, and it seemed like the perfect public setting for a deal to go down. As Thad approached the hostess, he was pleased and surprised to see that the restaurant was extremely crowded for six P.M. Then again, it was Orlando, which even in the summer was a haven for tourists from all over the world. Hell, he might as well have set the meeting for the middle of Disney. They could have exchanged contraband for cash on the way up Space Mountain.

Except, of course, there would be no exchange of contraband for cash until everyone felt comfortable with each other. Thad hadn't brought any of the samples along with him, and he was expecting the buyer to be just as cautious.

After the middle-aged couple moved out of the way, Thad walked up to the hostess and told her that he was meeting someone for dinner. He didn't give the hostess any names, nor could he actually describe whom he was meeting. In any event, the hostess told him that he was the first to arrive, so he opted to wait at the front of the restaurant.

A good five minutes went by as he watched at least a half dozen more tables get seated. The place was really bustling. Standing there, with so many people hovering around, he started to feel pretty nervous. Hell, he wasn't even certain that the other party was going to show up. Maybe the woman had chickened out at the last minute. Maybe she had even called 911. Thad knew his own nerves were working against him, and he had the sudden urge to just turn around and walk out of there.

And then he saw her, the woman as she had described herself in an e-mail, dark-haired, respectable-looking, wearing a tailored suit-skirt combination—she looked like a schoolteacher or a businesswoman, and there was a nervous smile on her youthful face.

She immediately recognized him from the outfit he had told her he'd be wearing: a black shirt and silver necklace sporting a dolphin pendant. The pendant had sentimental value—Sonya had given it to him years ago—but he wasn't sure why he had chosen it for this moment.

She shook his hand, introducing herself as Lynn Briley. Thad didn't give any name himself, and let the woman lead him, with the help of the hostess, to a four-seater against the right side of the restaurant, right up next to one of the curtained picture windows. Thad didn't see the woman's husband anywhere nearby, so he assumed that she had wanted to meet him first—which made sense, since she was the American. Kurt, Axel's brother, might not even have spoken English, for all Thad knew. Emmermann's e-mails had always seemed to be written in that very staccato manner of foreigners who'd learned English in school, rather than on the street.

After they were seated, and a waiter took their order—a random and hastily constructed list of Italian appetizers and entrées—they went right to business. Lynn had obviously noticed that Thad wasn't carrying anything with him; he was wearing shorts, sneakers, the shirt, and the necklace. So Thad wanted to quickly set her at ease.

"The samples are back at our hotel. After we're comfortable with each other, we can go back and exchange the money there. Does that sound good?"

She nodded, taking a sip of her water. She seemed as nervous as Thad felt, and that actually made him calm down a little. She was pretty, in that slightly older-woman sort of way, and he noticed that she had left the top button of her dress shirt undone, revealing the angle of her collarbone.

"Okay, where's your hotel?"

"The Sheraton."

"If you're more comfortable with that, we can do that. The Sheraton just down the street?"

"The big tall one," Thad responded. The woman was talking fast, and Thad really wanted to make her feel comfortable enough to relax. "On the left. It's been very nice. I'm telling you, this has been the most exciting event in my entire life, I think. Heck, I'm just hoping you don't have a wire on you! Anyway, you know what my girlfriend said today? She's like—they could make a life out of my movie."

Thad knew he was talking too much, but he couldn't help himself, he was starting to enjoy this, starting to really ride the adrenaline. The woman seemed to be easing up a bit also, and she seemed amused by his obvious enthusiasm.

"You sound very adventurous," she commented, "and your girlfriend must be very adventurous, too."

"What she meant to say is, they could make a movie of her life."

It was an extremely surreal comment to make—both for Rebecca, back at the hotel, and for Thad, here in the crowded restaurant, speaking to a woman who was about to pay him a hundred thousand dollars for stolen moon rocks.

Thad was starting to feel a bit more in control as he took a long sip of his own water. But there were still plenty of loose ends. He asked her about Kurt, her husband, and she explained that he was waiting nearby for her to call, to let him know that things were progressing. In return, Thad told her that his own partners were on their way to the restaurant and would be there soon.

"Do you want to talk to your husband before you meet the others?" he asked, wanting to move this along. His ten-minute grace period was almost up, and he expected Rebecca and Gordon to walk in at any moment.

The woman seemed to think about it for a second, then nodded.

"I tell you what. The music in here is really loud. Let me step out. I'll call my husband and get him on the way here, and while he's on his way, you can call your friends, and we can all sit down and chitchat. Sound good?"

Thad was about to answer, when he saw them—Gordon and Rebecca, strolling into the place as if they owned it, actually holding hands, although Thad suspected that Rebecca was just trying to keep Gordon from toppling over. Just as Thad had demanded, they took a table across the crowded place and called over a waiter. Gordon was talking extremely loud—so loud that Thad could hear him ordering Heinekens over the din of the other diners.

Hell, the guy was really making a scene—but it didn't seem like anyone else noticed, so Thad turned back to the woman.

"That's fine. I'll wait right here."

Thad realized he was sweating as he watched her go. All of his bravado from the moment before was gone, his nerves firing off, his entire being shaken by the sight of Gordon and Rebecca sitting there, across the way.

He took another sip of water, trying to compose himself.

. . .

Lynn Briley—aka Special Agent Lynn Billings—waited until she'd moved out of earshot of the suspect, whom she knew only as Orb Robinson, before pulling her cell phone out of her front pocket and placing it tight against her ear. She was breathing hard, though she wasn't particularly nervous; as an undercover agent with the FBI, she had conducted numerous missions in the past. Certainly, this was not the first time she had worn a wire, but there was always that special feeling you got when you strapped the electronics to your body—especially when you weren't certain what sort of environment you were getting yourself into. But Orb Robinson seemed pretty harmless. Of course, that didn't make him any less guilty.

"Kurt." She spoke rapidly into the cell phone. Even though she was out of earshot, she never broke character during a mission. "Things are going very well. His other two friends are here, um, that are involved in this as well. He does not have the samples on him right now, so they want to go to their hotel room. Which is the Sheraton just down the road by I-4. He said it's the big tall one. So his two friends are already in the restaurant. They have not sat down at the table yet. They want to get comfortable with you, and then we're gonna go and everything should be in the hotel room at the Sheraton; just advise—advise our friends. You know, why don't you do that right now. Um, well, go ahead and do that. I think that might help. Okay, and, um, just come in and we'll be waiting for you."

With that, she hung up and deftly slid the cell phone back into her suit pocket. She did a mental check, making sure the digital recording device was still well hidden beneath her clothes. Pasting a calm, collected smile back on her lips, she headed back to the table.

. . .

The woman was already talking, before she even fully settled back into her seat.

"I forgot to ask before I left," she said, and she seemed to be more relaxed after her phone call, "if you wanted him to bring the money. So, he's going to. Just, I figured that was the safest thing to do."

"Can we leave it in your car?" Thad didn't like this development at all. A suitcase full of money did not belong in this restaurant, and it seemed like an unnecessary danger. After all, they were going to all have to go back to the hotel anyway, to look at the moon rocks.

"You don't want him to bring it in?"

"No."

"It's up to you; I can tell him not to."

Thad took a breath. He didn't want to fuck this up by being too paranoid.

"Okay, I don't want to open up a briefcase full of money in here. But he can bring it in."

He wanted to keep the woman happy, and comfortable. Especially because he could really hear Gordon now, over the din of the restaurant, saying something to the waitress, something about some huge tip he'd obviously given her. Thad wasn't sure, but he thought Gordon was on at least his third Heineken. Which was kind of a terrifying thought, considering how high the kid already was.

"It doesn't matter," Thad quickly added. "I'll follow him out to the car afterward, and look inside real quick—"

He had barely gotten through the sentence when he saw a man approaching the table—tall, square-jawed, maybe a little too thin, wearing a somewhat stiff-looking blue blazer and a tie. Kurt Emmermann certainly looked European. And he was holding a briefcase in his left hand.

As he introduced himself, shaking Thad's hand and giving Lynn a little kiss on the cheek, Thad couldn't keep his gaze off that briefcase. Sure, he had no intention of opening it here in the restaurant, but he knew what was inside. *More money than he had ever seen in his life.* More money than he could imagine in one place. Enough money to change everything.

"Unbelievable," he said, realizing he was saying it out loud, but not really caring. "You spend so much time thinking about it. I mean, you see it in a movie in your mind, and then it happens. It's happening right now. It's weird. I almost feel like I've lived over the last two months, you know, this whole ordeal, I don't know how to feel. I really don't want too much more."

Both the woman and her husband were looking at him, maybe trying to decipher what he was saying, maybe just wondering what was going to happen next. The woman's eyes still seemed kind of amused, but the man was much more about business. Thad didn't care. He felt like he had one foot in the fantasy world he had been building for the past year, and one foot in reality. It was a wild sensation.

He quickly glanced at Rebecca, catching her eye. He was glad to see that Gordon was too busy with his Heineken to notice. Rebecca separated herself from the table and headed over by herself. She had to weave between a pair of diners being led by the hostess to their table—and as she passed them, Thad noticed something for the first time, something that seemed the littlest bit peculiar.

Other than the hostess, the other people in the restaurant—and there had to be at least fifty of them—all seemed to be middle-aged. No kids, no teenagers, no families. Nobody that was in their twenties, other than Thad, Rebecca, and Gordon.

Well, maybe there was some sort of convention nearby. Or maybe it was just Florida. Thad filed it away in the back of his mind. He stood as Rebecca reached the table, introducing her to Lynn and Kurt.

"You're really close to what I was expecting."

Thad wasn't sure why Rebecca had just said that, but from her voice, he could tell that she was really nervous. He gestured for her to take a seat, next to him. Kurt and Lynn were across from them. Lynn turned toward Thad.

"I don't even know your—Orb? I was going to say I don't even know your real name, but that's all right. You have one more friend coming, too?"

Thad shrugged, because he was really hoping that Gordon would just pass out at his table. Then, without warning, Kurt broke into the conversation for the first time, his words barely audible through a thick European accent.

"Now, this is exciting. I'm betting you will think about this for the rest of your life. You guys will be off to some beach somewhere, and you'll remember this day, this life-changing event. Very fun."

Thad glanced at Rebecca, who seemed to be put slightly at ease by the man's happy comments. Maybe she was picturing that beach. Thad looked toward the briefcase full of money, then back at the couple across from him. A ship in a storm that seemed to be settling, he'd pitched back to some level of confidence; he was ready to move this along.

"Well, we've talked quite a bit. I feel very comfortable. I think it's a good idea not to open the briefcase in the restaurant. And all of the samples are in the hotel."

He was squeezing Rebecca's hand under the table, and he felt her leaning into him, feeding off his renewed confidence.

"Oh, and tell them about the Antarctic meteorite," she burst in, her voice filled with energy.

Thad felt himself smiling. Rebecca was right. Why not have some fun with this? There was nothing he loved more than an enrapt audience.

"Have you heard of the ALH meteorite? It was collected in Antarctica. We have a NASA team that goes down every year. It's a great place to find meteorites."

The woman and her Belgian husband were leaning in over the table, obviously intrigued. Thad felt like he was back at the JSC, speaking to new co-ops, always the center of attention.

"Anyway, they bring them all back to NASA and start cataloging them. The first one they said, this looks so weird, so they called it 84, in the dilute form 001. They just put it in a big freezer, so it wouldn't be contaminated. They started studying it and noticed really strange stuff on it. It looked like microfossils. So they studied it more in depth and verifiably proved it's from Mars—"

And right in the middle of his lecture, suddenly there was Gordon, leaning in over the table, his breath stinking of alcohol. Before Thad or Rebecca could say anything, he was sliding into the seat next to them, his hand shooting out toward the couple across the table.

"Gordon," he said, by way of introduction. Kurt and Lynn shook his hand, and then he was shouting toward the waiter.

"Heineken!"

Thad felt his face getting red. But the couple seemed to take it all in stride, and they had already turned their attention back to him. He decided to just ignore Gordon as much as possible, and continued his story.

"So everyone also agrees that the stuff is actual microfossils. The question is, since it is here, what does that mean? Anyway, needless to say, it's one of the most famous rocks on the planet. I didn't put that in the list originally. Quite a find, huh?"

Lynn looked at her husband, then back at Thad.

"Isn't it in something to protect it, and keep it from being exposed?"

Thad nodded. She really seemed interested. Maybe she'd want to buy that, too. Maybe there were more briefcases full of money to be had. And maybe he would sell it, if the price was right. Hell, he was beginning to feel loose, like anything could happen.

"Yes, it's in a vial. Oh, you're gonna love it. This is like an extra bonus. We were very happy when we found it. At least, as a scientist, this—this one specimen is the most famous rock on the planet. Wow."

And then suddenly Gordon was butting in, his voice way too loud.

"Remember, you just saw something the other day about it on TV! He shows me it—remember this? It was on TV!"

Thad stared daggers at him. What the fuck was he even talking about? Gordon's eyes were totally bloodshot, his eyelids at half-mast. He was royally fucked up. But the couple still didn't seem to be bothered. The woman cleared her throat, drawing Thad's attention back to her.

"So, are you all mineralogists?"

"I'm actually into bioengineering," Rebecca butted in. "We go to school together. I'm biology."

"Are you really?"

"Bioengineering."

"Mineralogy, well, geology," Thad said, pointing to himself.

Gordon coughed.

"He's got three degrees he's working on. Yeah. He's a freaking genius."

Thad smiled thinly.

"I can pick between the three."

"The Cayman Islands!" Gordon suddenly shouted, taking them all by surprise. "Isn't that where we're trying to get anyway, with school? Sit on the beach and enjoy it! It's the Bible that's worked for us!"

Christ, he was really losing it. Thad glanced at Rebecca, and he could see she was thinking the same thing. They needed to wrap this up, and quickly. Thad lowered his voice, speaking directly to Lynn.

"Do you wanna just go ahead and get the check?"

Gordon butted in again before she could answer.

"I gave that girl back there a thirty-dollar tip!" he exclaimed. Thad wished he would just shut the fuck up, but Lynn seemed amused, rather than afraid.

"Are you kidding me? Dude, you're crazy."

"It was her first, I was her first table, ever. Aw, it was her first table. I said I'm gonna make your night. Belgium, eh, how is Europe now?"

This last was to Kurt. The guy seemed not quite to know what to make of Gordon, but he gamely tried to answer. For Thad, it was like watching a train wreck in progress.

"Europe is well. It's still Europe. It's home."

"Did Belgium go over to the euro?" Gordon shot back. What the hell was he going on about? "Your brother's Web site," he continued, now obviously drunk as well as stoned. "It was the first one. I'm like, wow, the guy just e-mailed me. He e-mailed me. Hey, did you tell him how many grams we have?"

Thad's teeth clenched as he glared back at Gordon.

"No, I think we're gonna get into that later. Just relax."

And he quickly made a signal to the waiter, who hurried over. Lynn reached for the check, stopping Thad before he could offer to pay. Gordon seemed to find this immensely amusing.

"Wow, you're competing now. You can always win on that one."

And that was all Thad could take. He stood, gesturing for the rest of them to do the same. Lynn put a couple of big bills down on top of the check, indicating that she didn't need to wait for the change. And then they were all heading across the restaurant toward the front door. Lynn suggested that Thad ride with her and Kurt, and that Rebecca and Gordon follow behind in the other car.

"Yeah, that's fine," Thad responded, liking the idea of separating the couple from Gordon as much as possible. "Um, which room are we in again?"

"We'll meet you in the lobby," Rebecca quickly responded.

"We're going to the Sheraton at I-4, right?" Lynn interjected.

"Yes," Rebecca said, hitting the door first. Gordon was staggering behind her, but he made it through the open doorway without losing his footing.

"Wait for us in the lobby," he called back, slurring his words, "if you get there before us."

Thad nodded, but Gordon and Rebecca were already out of the restaurant and hurrying across the parking lot toward her car. He looked at Lynn, who smiled amiably back. Kurt was already reaching for his car keys; his other hand still gripped the briefcase, which swung heavily by his left thigh as he exited through the open doorway. Somehow, its rhythmic, pendular motion helped quiet the thoughts racing through Thad's head. Gordon was out of control—but the situation wasn't. Actually, things seemed to be going very smoothly.

One hundred thousand dollars, one short car ride away.

As the woman held the door open for Thad, he smiled at her. She smiled back—but in that brief second, he noticed that she was actually glancing past him, at something in the parking lot. He quickly followed her gaze—but it was just another couple of restaurant patrons getting into their own vehicle on the other side of the lot. A man and a woman, actually, dressed pretty formally for a warm Saturday evening. And they both appeared to be in their late thirties or early forties.

Odd—but Thad pushed the thought away. He told himself again, he was just being paranoid. In a few minutes, they would be back at the hotel. And then it would be just him, Rebecca—and a briefcase full of cash. After that—maybe there would be a nice, pretty beach, with plenty of palm trees to go around.

35

Thad was still thinking about that perfect, pretty beach as they pulled into the Sheraton parking lot, Rebecca and Gordon a single car length behind them. Lynn and Kurt had been pretty talkative for most of the short drive over from the restaurant, shifting through a range of topics, from the muggy weather in Orlando to the best beer makers in Belgium—and pretty much everything in between. Thad was starting to really like them, and even found himself wondering if they'd all stay in touch after the deal was completed. He was certain that Gordon would be out of the picture as soon as he got his ten grand, but Thad and Rebecca would one day want to travel to Europe—and it would be nice to have people there to show them around. Kurt could introduce them to his brother, and Thad could finally meet the man behind all those e-mails. He was sure he'd have a lot in common with such a conscientious rock hound. Hell, maybe they'd all end up visiting that pristine beach together, share some laughs about the deal that had brought them together.

But the minute Lynn jammed her foot on the brake, sending the car skidding to a sudden, screeching halt, Thad's mind went absolutely blank, the imagined beach swallowed in a burst of pure and instant terror. He opened his mouth to say something, but before he could find any words, Lynn and Kurt were out of the car—and then there were men racing at him from every direction, shouting and screaming and pointing—

And then Thad saw that the men had guns. Dozens of them, everywhere, all over the parking lot, guns of varying sizes, pistols and automatics and even things that looked like sniper rifles, all of them drawn and aiming right at his face. Bright light exploded everywhere at once, illuminating the entire front of the hotel. Thad gasped, pressing back against the car seat, trying to disappear into the sticky, sweaty vinyl. But then one of the men was grabbing at the car door, and suddenly there were hands all over him, yanking at his shirt and his hair and even his skin. As he was dragged out of the car—above the shouting and the screaming—he could hear the *thump thump thump* of a helicopter up above. The shadow of the thing passed right over him, the fierce wind from the rotors pulling at his hair—and then it was past, out over the highway. And Thad could see, beneath the copter, at least twenty police cars, lights flashing, parked behind barricades and yellow tape. They had closed International Drive; in fact, it looked like they had shut down an entire section of the city.

"On your knees!" screamed a voice next to his ear. "Now!"

It was Kurt, but now Kurt wasn't talking about idyllic beaches, and he didn't have a Belgian accent. Now Kurt was aiming a .32-caliber handgun at the back of Thad's head. And there, just a few yards away, was Lynn, but she wasn't asking about his adventurous girlfriend or the movie of his life. Now there was a badge affixed to her suit jacket, and she was talking to two men in police uniforms—and they were all looking at Thad, and one of them was smiling, but it wasn't an amiable smile; it was a mean, arrogant kind of grin.

And Thad knew, with every fiber of his being, that he was fucked.

He felt the tip of a shoe kicking at the back of his legs, and then his knees hit the pavement. A heavy weight pressed against the small of his back, and then he was down flat, his left arm being pulled behind his back. He could hear the clink of handcuffs being readied, and in that brief moment he felt his entire life energy flowing out of him, like a cork had been pulled out of his heel and all of his dreams and accomplishments and beliefs were just running out of him, water from a pierced balloon. And he knew, right then, that this was a perfect time to die. Up until that point, that very second, everything in his life had been so incredible and exciting. He was a NASA scientist with a chance of one day becoming an astronaut. He had a beautiful girlfriend and a beautiful, though separated, wife. He could speak multiple languages and fly airplanes and cliff-dive and swim in the NBL. He had ridden in the Space Shuttle Simulator. He had everything.

And now it was all gone, poof, everything he had ever worked for, everything he had ever achieved. Gone.

He knew immediately what he had to do. He glanced up, and even from that angle he could still see all those guns aiming at his head. Thirty, maybe forty of them, Christ, even though, of course, they knew he was unarmed, he was wearing shorts and a shirt and had just spent the past hour in a restaurant talking about moon rocks and Mars meteors. Forty guns, more than enough to do the job. The handcuffs weren't locked on yet, he had a second left before it was too late—all he needed to do was roll over and start swinging. Hit one of the cops or the feds or even Kurt in the face, get them to start shooting. Thad wouldn't even feel a thing.

But then, out of the corner of his eye, he saw the commotion across the parking lot—Gordon and Rebecca being dragged down to the pavement just like he was, another dozen or so cops swarming over them like maggots over meat. Gordon was one thing, poor sap had screwed himself by coming down to Florida—but Rebecca . . . Christ, Rebecca. He could just barely see her tiny form splayed out on the pavement, her wrists being pinned behind her lower back.

Tears burned at the corners of his eyes. Rebecca. He had to help her. He had to make sure she got out of this okay. He had to protect her. And if he died here, in this parking lot, she'd end up in prison, maybe even hating him for the rest of her life. He couldn't let that happen. He had to live, to make sure she continued to love him. To make sure she stayed safe.

He let the last few drops of his life energy dribble out the bottom of his heel, closed his eyes—and listened for the piercing, metallic crack of the handcuffs clicking tight around his wrists.

36

Axel Emmermann didn't truly understand the enormity of the situation—or the storm that was headed his way—until he saw the look on his fifteen-year-old son's face. Sven had come through the door to Axel's bedroom at a full run, and now he was just standing there, eyes wide, cheeks flushed, as he struggled to catch his breath. Christel was already out of bed and on her feet, rushing toward the boy to see if he'd somehow injured himself, but Axel waved her back, having a pretty good idea that Sven's nearly catatonic state had something to do with the flurry of phone calls Axel had received via his cell phone the night before.

That, in itself, had been unusual, because he almost never used his cell phone, and he'd missed the first few calls trying to find the damn thing. When he'd finally discovered it in the bottom drawer of his dresser, it was still ringing; he'd been surprised to hear the familiar voice of the president of the Antwerp Mineral Club on the line.

His old friend had sounded as breathless as Sven now looked. The president himself had just gotten off his own phone, having received a panicked call from his elderly mother. The woman, deep into her eighties, had been a sort of mascot to the Antwerp Mineral Club for some time. Apparently, a journalist who had once written an amusing piece about her interest in rare rocks had tracked her down in the middle of the night. Because of her age, she had been pretty confused by the call, and had simply passed the journalist's information to her son.

"Something crazy is going on," the president had gasped, once he'd gotten Axel on the line. "There's been some sort of major arrest in the United States, and it seems that it somehow involves our mineral club."

Axel had nearly dropped the cell phone. He hadn't heard anything from the FBI or Orb Robinson in over a week. He had dutifully passed the baton to the people who were supposed to know what to do with it, and even his wife had finally let the issue drop. The last thing he had expected was to hear about it again—via the eighty-year-old mother of the president of the Antwerp Mineral Club!

But apparently, the Belgian journalist had been eager to hunt her down because the FBI had issued a little press release. Buried deep within that release was the mention of a Belgian collector from the Antwerp Mineral Club. A reporter from Tampa, Florida, had contacted a colleague from Belgium—and the trail had led all the way back to Axel Emmermann.

The president of the club had guessed, correctly, that Axel hadn't simply erased Orb Robinson's e-mail as everyone else had done. He had taken it upon himself to do something about what they all had assumed was a hoax. After the president's initial shock had worn off, he had become very excited at the prospect of all the coming press. Axel's actions had put the Antwerp Mineral Club on the map.

Axel's thoughts had been swirling as he'd hung up the phone, but he hadn't even had time to inform Christel when the phone was ringing again. It was the *Tampa Herald*, a newspaper all the way on the other side of the world, calling him to talk about his role in bringing down Robinson. The reporter hadn't gleaned much information yet—just that arrests had been made, and that the people arrested were connected to NASA. Nevertheless, the journalist treated Axel like a hero. And at the end of the conversation, the man warned Axel that this was probably just the beginning. A crime this big had never happened at NASA before. There was a good chance it would become an international story.

Looking at Sven's face as the poor kid stood in the doorway to the bedroom, Axel had a feeling that the journalist had been correct.

"There's something being erected outside my bedroom window," Sven finally managed. "It looks like it might be some sort of spaceship."

Axel looked at his wife, then quickly rushed out of the bedroom. He barreled down the hallway to his son's room. Christel and Sven were right behind him, his wife holding on to the back of his shirt as they went. When he got to his son's window, he yanked back the drapes—and Christel gasped behind him.

Rising up on his front lawn was a giant steel television antenna. Behind the antenna, there were at least two news trucks with satellite dishes affixed to their roofs. There were reporters everywhere, a few he even recognized from the local nightly news. Beyond the trucks, he could see his neighbors gathering outside on the street, even though it was barely five-thirty in the morning.

Axel turned and grinned at his wife. He didn't need to say anything, because he could see from the expression on her face that she was equal parts stunned and proud.

Axel was now an *international* superhero.

37

The holding cell was in Tampa, a bit of a drive via police caravan—lights flashing, sirens wailing—from Orlando, but it might as well have been on Mars; everything had become so surreal and foreign and confusing, and Thad had no choice but to just go with it, handcuffed and eventually shackled, a metal chain running between his wrists and his feet, fingerprinted and shoved along by a never-ending parade of police officers and FBI agents and people with badges he couldn't even recognize. By the time he finally was led into the holding area, he'd been interrogated at least twice, but he had remained utterly silent—more the result of his stunned state of mind than from any sense of strategy. But the minute he saw Rebecca in the holding area, separated from him by the bars of their individual cells, his mind cleared, his senses sharpened. The world snapped into focus like a leather belt pulled tight, and he was able to zone out the dozens of strange and terrifying people staggering around the huge, open tank right next to his isolated cell—most of whom looked drunk and high and crazy, a few shirtless and even one completely naked, the smell of feces and sweat and fear so thick it made Thad want to gag. Instead, he focused on Rebecca, only Rebecca.

Her face was as white as the lightest part of the moon, and there were tears streaking down her cheeks. She was curled up in a neat ball, right up against the bars, so close that Thad could almost reach out and touch her. She saw him, but she didn't even unfurl herself; she remained a little fetal ball, her shoulders rocking with each sob.

"It's going to be okay," he said, mustering enough strength to make his voice steady. "I promise, you're not going to take any of the blame for this. Just tell them you don't know anything."

"I already told them everything!" she half wailed, and Thad was momentarily taken aback by the viciousness in her voice. She was beyond terrified, desperate, and devastated. "And they made me call my parents."

"It doesn't matter what you told them," Thad said, not sure whether it was true, but trying to regain control, even in the most uncontrollable situation. "I'm going to take the blame for everything. You tell them you were afraid, that I forced you to do this, that you didn't know anything about the moon rocks. I need you on the outside. I need you free, so that I can talk to you, so that you can be my lifeline."

And he meant it. He knew that he was going to prison. The only way he'd be able to survive was if she was outside, free, living her life, and communicating with him. He believed that he could get through anything, as long as he could talk to her once in a while, hear her voice, tell her he loved her.

"But my dad—he's coming to get me. And he says I can never talk to you again."

This hit Thad harder than anything else so far. He shook his head.

"No, we have to stay in communication."

"They told me we could get thirty years. Thad, I can't go to jail for thirty years."

Thad lowered himself to the floor, his head against the bars. He wished he could reach out and touch her. But she was too far away. *Thirty years?* It was probably bullshit. It had to be bullshit. What they'd done—it was a prank, a mental game that had gone a little too far. Shit, the cops were just trying to scare her. And they'd done a good job.

"You're not going to jail; I'm going to tell them that it was all my idea."

She raised her head from her hands, and the sobs seemed to subside a bit. Maybe his words had finally gotten through to her.

"But my dad—"

"For now, just do what he says. After a little bit, when you're out of here, we'll find a way—"

But he never had a chance to finish, because suddenly there were uniformed officers at Rebecca's cell, and they were telling her to get up and follow them. She threw one terrified final look at Thad, and then her head was down, almost to her chest, she was moving quickly in the direction that the police officers had indicated—and a moment later she was gone, and Thad was alone. He breathed deeply, trying to catch one last whiff of her floral perfume in the air, even the tiniest molecule of her passing to keep him from completely coming apart—but there was nothing there but the fetid stench of that Floridian purgatory.

It was his turn to curl up into a fetal ball, his mind going numb.

. . .

"One phone call. You have five minutes."

Thad stood in front of the pay phone as the uniformed officer stepped away, giving him a few feet of privacy. The bulky hunk of metal and plastic hanging from the wall seemed so utterly anachronistic, and Thad couldn't help but remind himself that just two days ago he had been listening to voices through a bone-conducting receiver, and now here he was, standing in front of an ancient-looking pay phone, the sound of drunken gangbangers and transients echoing all around him.

Thad had no idea how long he had been in the holding cell before the officer had come to get him for his legally sanctioned call. He had considered telling the man just to leave him alone; the phone call wasn't going to do Thad much good, because he only knew one phone number, and the person on the other end of that number wasn't going to be very helpful.

The minute Sonya's voice echoed over the line, he knew that his prediction was correct. She was furious. He was in jail, calling her collect—

and her fury only grew as he gave her the details of his situation. Not only had he gone through with this heist, but also he had done it with Rebecca, a girl he had known for less than a month. In the course of the short phone call, Thad realized that Sonya still had strong feelings for him—that somehow, even though they had barely spoken over the past few months, she had still harbored the thought that someday they would work things out.

Thad had put an end to that. The robbery, even jail time—these were things Sonya could have gotten past. But that he had done this with a girl other than herself—that was unforgivable.

He tried to talk past her anger—because he needed her help. He had been told by one of the federal courthouse officials that he was going to be given a signature bond, which meant that any adult in the country who didn't have a criminal record could go to any courthouse and sign him out, to await his trial. It didn't need to be a brother, it didn't need to be a parent, it didn't even need to be an angry ex. Just a signature from any adult, and he would be free until they were ready to try him for the crime.

"If you won't do it," he pleaded into the phone, "if your parents won't let you or if you just can't because you need to move on—I understand. But please, reach out to the people who know me, to anybody you can think of. Maybe someone at NASA, maybe someone from school—"

But Sonya shut him down with one of the harshest things he had heard since his parents had disowned him.

"There isn't anybody. Nobody is going to take responsibility for you now. Don't you realize what you've done?"

Thad stood there, frozen, trying to think of something to say—when the call cut off.

Thad wanted to call her back, if only to tell her that he was sorry for everything. But even if the police officer hadn't already started prodding him back toward the holding cell, he doubted that Sonya would have accepted the call.

Thad realized that nobody was going to come get him. Even such a simple thing as signing a piece of paper—nobody was going to come for him. Rebecca couldn't because of her father. Sonya wouldn't because she was angry, maybe because she was scared—of what her family would do, of what it would mean to her prospects of moving away from a failed relationship. And beyond them, there was no one else. Thad didn't have parents anymore. And his friends—at NASA he was now a pariah. The other co-ops would avoid him like the plague. The esteemed scientists—he had been an amusement to them, a promising kid who told adventure stories for their entertainment, but that was all.

He was alone; he was in the system. And no one on Earth was going to help him now.

• • •

They called it the Submarine.

The county jail on Orient Road in Tampa was the most miserable place Thad had ever seen. Just hours after his phone call to Sonya, he was led down a stairwell, through an endless parade of iron doors and barred windows, into a long cement hallway, bordered on both sides by tiny metal windows covered with steel, all of it painted an unnerving shade of blue. He was handcuffed and shackled and wearing a belly chain, shuffling along with his head down, prodded from behind on both sides by uniformed officers. He was doing his best to keep his mind completely blank, because any thoughts that could erupt in a place like this would do him no good. He had to become an empty shell, because he knew that he was going to be here a long time.

About a quarter of the way down the hallway, the guards stopped, and one of them stuck a huge metal key—about four inches long, like something out of a medieval dungeon—into a panel, unlocking a steel door. There were four pump levers on one side of the door, and it took two guards to turn them, forcing the heavy slab of steel to slide open, inch by inch. They paused when there was just enough room for Thad to be shoved inside; first they unlocked his handcuffs, undid his shackles;

then one of the guards gave him a sarcastic little pat on the back. Once he was through, the door was slammed shut behind him.

Directly ahead was what was called the dayroom, to the left the bedroom. Thad took it all in with quick flicks of his eyes. In the bedroom he saw eight bunk beds, basically steel plates bolted together, attached to industrial-looking iron frames, two rows of four. Standing between the bedroom and the dayroom, he saw two steel toilet seats—no lids, just the toilets themselves, standing there in the middle of the open area in full view of everyone—and a single shower off in one corner. On the far end of the room, opposite the hallway he had just walked down, was what they called the catwalk. It was just a bunch of bars separating the dayroom and bedroom from another long hallway, where guards took turns walking first one direction, then the other. There would be no privacy of any kind.

As Thad took a little step into the dayroom, his stomach tightened into knots. In the middle of the room stood a couple of metal picnic tables, with bolted-down benches. There were two pay phones against one wall, both currently occupied, and beyond them, affixed to a joint a few inches from the ceiling, a television set. There was a knob on the TV, but even from that distance he could see that there were only two numbers on the knob, two stations available. At the moment, the television was on, and Thad recognized a children's show—something called *Teletubbies*.

At the picnic table farthest from Thad, a group of three African American men in bright orange prison overalls were intently watching the show, every now and then bursting out in a concert of what sounded like truly crazed laughter. The men looked just like Thad would've expected—angry, tattooed, overly muscled, and terrifying.

At the other picnic table, there were two white men; one was huge, maybe three hundred pounds, his gut hanging out over his orange pants. The other was half his size, with a goatee and an enormous tattoo running up the left side of his neck. There was a deck of cards on the table in front of them, and the larger man was in the process of throwing down a card. A low number came out, and this seemed to be a good thing, because the man laughed and clapped his hands against the table. Then the smaller man took the next card, threw it on the table—showing a king. The man snarled, then leaped off his bench, got down on the floor, and did ten push-ups.

As Thad watched the two tables of men, his entire body started to shake. He couldn't believe that this was now his life. Three days ago, he had been diving in the NBL, he had been hanging out with astronauts, shooting the breeze with some of the smartest men in the world. And now he was in hell.

Before he could take another step into the room, one of the black men from the *Teletubbies* table crossed toward him—swaggering like his feet weighed a hundred pounds each. He had muscles everywhere, and there was a hardness in his face that sent chills into Thad's bones.

He stopped a few feet in front of Thad, looking him over. Then he grinned, his teeth a peculiar shade of yellow.

"My name is Graveyard. Graveyard Serious."

He gave Thad a hard punch to the shoulder. Thad did his best not to flinch. The man turned and headed back to his *Teletubbies*.

Thad stood there, waiting, but nobody else acknowledged him—so he quietly crossed into the bedroom and made his way to what appeared to be an empty steel bunk. As he lowered himself onto the bunk, he realized that it was a solid sheet of metal, with tiny holes drilled into it that were supposed to make it the littlest bit flexible. No mattress, no sheet. There was, however, a pillowcase—to remind Thad that he didn't have a pillow.

He lay down on the bunk, wrapping the pillowcase over his eyes. He could still see the bright lights, even through the material of the pillowcase, and there was a loud buzzing coming from the fluorescent panels. He knew he'd never be able to fall asleep. Instead, he tightly shut his eyes and started to cry.

. . .

"Houston, we have a problem. Houston, we have a problem."

Thad's eyes tore open as the words reverberated through his ears, and he jerked himself up into a sitting position—nearly slamming his head on the steel bunk above him. It took him a minute to recognize his surroundings—to realize that it hadn't all been a dream, that he wasn't lying in his apartment back at NASA or curled up next to Rebecca in the parking lot of a Baptist church. He was on the bottom bunk in a jail cell, wearing an orange jumpsuit, with an empty pillowcase wrapped around his eyes. There were at least seven other men in the room with him, in various phases of sleep—even though the place was still lit up as bright as day by the ever-buzzing fluorescent ceiling panels.

"Houston, we have a problem."

It took Thad another moment to realize that the words were not in his head, that they were actually reverberating around the entire cell—through the entire county jail, actually—via the guards' intercom system.

"Houston, we have a problem."

This time, the words were followed by a moment of wicked laughter; whoever was speaking into the intercom was having a grand old time. Thad looked around, trying to figure out what the hell was going on. And then he saw the muscled black man approaching his bunk.

It was the guy who had introduced himself as "Graveyard Serious," and he was holding something in his left hand. For a brief second, Thad's mind whirled through every prison movie he'd ever seen, and he fully expected a steel shank to be driven through his throat. But instead, Graveyard tossed the item at his chest, where it landed with a soft thud. It wasn't a shank. It was a newspaper. And the banner headline across the top of the front page was all about Thad.

"'Moon Rock Heist,'" Thad whispered, reading the words as he saw them.

He looked up and saw that the other prisoners were all out of their bunks now, gathering around him. Graveyard was pointing a long finger at the newspaper.

"One of the hacks gave me that. You're the fucking talk of Orient County. Boys, we've got ourselves a celebrity."

Thad felt his cheeks flush as he read the article. It was all there, in black and white. The arrest of Thad, Gordon, and Rebecca. And Sandra—according to the article, she had been arrested, too, dragged right out of her job in handcuffs. The newspaper was calling it the most significant heist in NASA's history. Everett Gibson, whose lab had been robbed, had actually been taken in for questioning upon returning from a trip to Australia.

Before Thad could read any deeper into the article, Graveyard grabbed the newspaper out of his hands and waved it in front of the other prisoners.

"That's right, say hello to Moon Rock!"

And just like that, the name stuck. *Moon Rock.* Thad laid his head back against the hard steel bunk as the intercom continued to bray in his ears.

"Houston, we have a problem."

38

"Moon Rock, you're up."

Thad was only on his seventh push-up, and he owed Graveyard three more—but the guard standing by the open cage door looked serious, and even Graveyard wouldn't have ignored a hack's offer to get out of the claustrophobic cell, even if the reason was still completely unknown.

Thad pulled himself to his feet and pointed at the face card that was on the picnic table between him and the other prisoner.

"I'll finish my ten when I get back."

"*If* you get back, Moon Rock. Maybe they're about to let you go."

Graveyard bared his yellow teeth, amused by his own statement. It was still only a few hours into the second day, so there was no chance in hell that Thad was going anywhere. But he was happy to get out of that cell, even for a moment. None of the inmates had made any attempt to kill him yet, but there was such an undercurrent of anger and subverted violence in that place; it probably had something to do with the shared, open toilets, or the incessant caterwauling of the Teletubbies. The jail was so infused with bad feelings, Thad would have done just about anything to get out of there.

He crossed to the door and held out his arms for the proffered handcuffs. After the cuffs came the shackles, and then the guard led him down the long hall. The next thing he knew, he was being brought into a small room with cement walls and no windows. There was a steel table in the middle of the room, and four metal chairs. Thad was handcuffed to one of the chairs, then left alone with his frightened thoughts.

Five minutes later, Thad's court-appointed attorney entered the room, followed by two women. One of the women identified herself as the OIG—a federal officer from the office of the inspector general, attached to NASA. The other woman—with severe-looking eyes and a tight bun of brown hair—introduced herself as the prosecutor assigned to Thad's case.

The truth was, the two women were about as familiar to Thad as his lawyer. The man in a stiff blue suit had been little more than a name on a sheet of paper Thad had been asked to sign when he'd first been checked into the county jail. His first name was John, and to Thad, he seemed like he was just out of law school. Maybe he would one day be a wonderful lawyer, but Thad had the feeling that at the moment, he was just trying to get through the day.

As the three of them took their chairs, Thad began feeling incredibly self-conscious. He was still handcuffed, chained up like he was going to kill somebody, like he was this dangerous criminal—and not a NASA scientist who had done something stupid. Before he could say anything, his lawyer placed a tape recorder in the center of the table and started asking questions. About the heist, about the planning, about Gordon and Rebecca and Sandra, about Everett Gibson and the moon rocks—about everything. He was doing all of this right in front of the prosecutor and the federal officer, and Thad just stared at him, trying to figure out what the hell was going on, trying to understand if this was how it was supposed to work.

When it became time for him to answer, Thad shook his head, giving his lawyer a plaintive look. The man seemed to understand, and he quickly asked the two women to give them a moment alone.

After the women had left, shutting the door behind them, the lawyer started over. He explained to Thad that NASA, the prosecutor, and the FBI had a lot of questions they wanted answered—and there was a chance that because of this, Thad would be able to make some sort of deal. NASA wanted to know exactly how the heist had happened: how Thad had been able to get inside Gibson's lab, how he had known about the moon rocks—everything that wasn't already on tape from the sting operation at the restaurant. And most important of all—Everett Gibson had told the FBI that the safe had contained his life's work, a number of green notebooks that were filled with thirty years of his scientific research. He had intended to use those notebooks to write a book after he retired, and they were considered invaluable.

Thad shook his head, his mind whirling. He didn't remember seeing any green notebooks in the safe. As far as he knew, they hadn't thrown anything out, other than the safe itself, so if there were notebooks, they'd still be either in Sandra's storage shed or in the suitcase that had been with them in the Sheraton. But Thad didn't really want to talk about some phantom notebooks; he wanted to talk about Rebecca.

He wanted to know what was going to happen to her. His lawyer seemed shocked that this would be Thad's first concern—but he did his best to explain the situation. He said the way the system worked, there was a certain amount of mandatory time a judge could give someone for taking part in a crime like this—based mostly on the value of the stolen items, since there hadn't been any acts of violence committed. That value was still to be determined, and much of any trial would be about figuring out exactly how much 101.5 grams of moon rock was really worth.

Each level upward from the minimum sentence was called a "departure." For the crime Thad and his friends had committed, they were currently looking at a maximum of three departures—or, roughly, three years in prison.

Thad's stomach dropped as he heard those words—*three years.* Picturing the cell he had just come from, the open toilets, the steel bunk beds, the guards and the inmates—he couldn't imagine how he would survive that. Then he pictured Rebecca—there had to be something else, something he could do.

Thad's lawyer admitted that there was, in fact, another way—for the girls, Rebecca and Sandra, at least. They could claim that they had been coerced into the crime, and had taken a minor role, led astray by a criminally intent "leader." John guessed that this was something their lawyers would probably advise them to do—but he assured Thad that this was something he would fight, tooth and nail, because if Thad took the role of the leader, he would be looking at an even longer sentence.

Thad stopped him right there, his handcuffs clanging together as he tried to raise his hands. That was exactly what he wanted to happen. Not the longer-sentence part—but for Rebecca, he would accept the role of the leader, if that meant Rebecca could remain free. If she stayed free, she would be his lifeline. He didn't have anyone else.

The lawyer looked at him, rubbing a hand over his tired eyes. He asked again if Thad was sure that this was what he wanted to do—basically lie down and let the girls argue that they were coerced or seduced into this crime. Thad nodded. The lawyer asked a third time—reminding Thad that from what Rebecca had told the FBI, Thad had only known this girl for a month. He was willing to throw away years of his life for someone he had known only four weeks?

Thad nodded again. He didn't think of Rebecca as someone he had only known for a month. She had filled something inside of him, something he'd needed; whether that was something his mind had invented or something real—it didn't matter. Her life had to continue.

Finally, the lawyer shrugged. Thad was his client, appointed by the court. He wasn't a friend or a family member. It was Thad's life. If Thad cooperated with NASA and the FBI, they would maybe go lenient on him, but if he was the leader, the self-admitted ringleader—well, he was looking at three years in federal prison, maybe even more.

Thad nodded, willing his brain to ignore the thought of all those years—and told the lawyer that he would do what he had to. The man shrugged again, and signaled the guards to send the prosecutor and the federal officer back into the room.

I know that you will never read these words, but I still need to write them down. I need some way of expressing your effect on me. I need to shape the tears into words. You once asked me why I love you . . . a question that has no answer on this side of the horizon. I can no more explain "why" than I can explain why I am self-aware. Every thought I have, every sensation and emotion comes laced with the knowledge that I love you, that I desire you, that I long to know your happiness, but questioning why gets beneath the question of my very existence. Still there is another question that you deserve an answer to—the question of "what" I love about you. To be fair, this question is also impossible to answer, but only because it is impossible to exhaust. Each brushstroke, however, belongs to the same painting, every detail reflects the whole.

39

Thad had always been a quick study.

At NASA, being quick to pick up how things worked had been important because it had caught the attention of the people Thad needed to impress, and it had given him that extra edge so that he could construct the person he wanted to be, right from day one.

In county jail, being quick to pick up how things worked was important because it kept Thad alive. Not only in that clichéd, late-night prison-movie sense—although there was always the very real risk of looking at someone the wrong way, saying the wrong thing, getting inadvertently involved in something that could easily have gotten him killed—but also in the sense that if he wasn't able to get his head around the new reality of his life, he was going to be lost in a place where even his fantasies couldn't protect him.

It was conventional jailhouse wisdom that it took about two years for a man to reach empty, to finally let go of his old life—hopes, dreams, expectations, family, real contact with the outside world—two years to reset at rock bottom, to become that empty, unimprinted shell. By the end of his first year of being locked up, awaiting sentencing, Thad knew that the jailhouse wisdom was probably correct. He was halfway to becoming that nowhere, nothing man, and if he had to endure another year, the time would shatter him and cause him to shed whatever was left of his old self.

The worst moment of each day usually came when he lay down on his hard steel bunk, listening to the incessant buzz from the brightly lit ceiling, waiting for the *clump clump clump* of the hacks' boots as they walked along the catwalk, often trying to ignore the horrifying, muted groans of men in nearby cells being abused, beaten, sometimes even raped by other inmates. It was a half-awake, half-asleep kind of place, where it was impossible to shut down his senses but equally impossible to digest what he was seeing, hearing, smelling.

The best time of the day was when he found himself alone in the shower, because it was the only time he could let go and cry.

In between, there were moments, good and bad, that marked the monotony of life in a cage. Meals, almost always grits, served on plastic trays that had to be returned and counted. Exercise, in a yard barely fit for a dog, fetid and hot and dangerous, where Thad usually stood in a corner trying not to catch the attention of anyone who might do him harm. TV time, usually those damn *Teletubbies*, sometimes the news, other times a Christian station spouting Scripture. And then, the card games with his cell mates—during which Thad was often asked to retell the story of the Moon Rock Heist—which inevitably morphed into a discussion of the sort of sentence he was probably going to receive, now that he had pleaded guilty and cooperated with the FBI.

Like everything else in prison, the topic of his sentence had become something the prisoners were eager to gamble on; not just Thad's cell, but all of the surrounding pods got involved, inmates choosing sentences they thought Thad would receive; anyone who missed by more than a year was going to have to do fifty push-ups, one of the few forms of currency allowed in the jail.

Although Thad's lawyer was still convinced that the highest penalty that Thad could receive—no matter how much NASA and the court's experts finally decided that 101.5 grams of moon rock and the little Martian meteorite were worth—was about three years, a handful of prisoners had guessed as high as five. Thad knew that there was no way he could survive being caged up that long, but even so, he never once regretted pleading guilty, or disallowing his lawyer to argue against his being in the leadership role of the heist.

His shouldering that weight had allowed Rebecca and Sandra to plead that they had been misled, coerced, and taken minor roles in the theft. When Rebecca's sentencing day finally came—a year after the heist—Thad was engulfed by a mixture of feelings. He hadn't had any contact with her since the day of their arrest, and every passing minute without that contact had been sheer torture. Every time he'd spoken to his lawyer—his only real link to the outside world—he had begged the man to get him in touch with her, to give him a phone number, an address, anything, but the lawyer had explained that it was impossible. Rebecca had been preparing for her own day in court—and as she had said, her father had banned her from speaking to Thad ever again.

But now that she was getting a sentence, Thad allowed himself to hope that afterward, things might change. When he found out that she had received only probation, along with 180 days of house arrest—he was thrilled. She wasn't going to jail, she was free, and eventually, he believed, she would reach out to him. Sandra, too, had gotten probation and house arrest, having also argued a minor, coerced role in the plan. Thad had been painted as a charismatic Svengali, a good-looking, fast-talking lothario who had duped the poor innocent girls into following him into Everett Gibson's lab, but he didn't care what they said about him because it had gotten Rebecca off, and she wouldn't have to go through what he was going through.

Gordon hadn't been so lucky, but it had been the stoner's own fault. He hadn't shown up for his court date, had instead gone on the run. When they had finally tracked him down in a Utah state park, he had stayed true to form—giving his name as Job, from the Bible, ensuring that the wrath of an angry government was going to rain down on him come sentencing time.

But Rebecca was free—and yet, Thad still had no way of reaching her. Over the course of the next few weeks, it became an obsession—and he began to try finding ways to contact her, if only to hear her voice one last time. Every time he heard of a prisoner being released, he'd approach the man, begging that once the man was on the outside, could he look up a girl named Rebecca Moore, and send Thad what he found? Most of the inmates looked at him like he was crazy, some openly laughing at the idea that they would have any more contact with the jail once they were out that door.

Realizing he wasn't going to make any progress that way, Thad created a game to try to achieve the same results. Using a piece of newspaper that one of the inmates had gotten from a guard, he re-created a puzzle he had learned back at NASA—actually, in a study aid designed to help potential astronaut applicants, as it was a test often given during the astronaut application procedure. As the other inmates watched, Thad tore the sheet of newspaper into five geometric shapes. These shapes, he explained, could be rearranged into a perfect square—but there was only one way to arrange them so that they fit together as a square, and there would be a time limit involved. Thad knew that NASA applicants usually took about ten minutes to get the arrangement correct. So he gave the inmates twenty, betting them a meal on the result. If they could create the square in under twenty minutes, they would earn Thad's dinner. If they lost, their dinner was Thad's.

One after another, the inmates failed; each time, Thad traded back the won dinner for a single request—find Rebecca, and tell her that Thad Roberts loves her. That was it, not even an address or a phone number—just tell her that Thad still loves her.

But even as a year dragged into fourteen months, Thad never received any indication that Rebecca had been contacted. No mail from ex-inmates, not even a postcard. The only mail he did receive came from Sonya, in fact. Divorce papers, with a blank spot where he was supposed to sign to make the separation simple and final for her, so she could move on with her life.

Thad didn't have to think about it for very long; it was the least he could do, and he knew that Sonya deserved to be happy, and to forget about him. Since he did have her phone number, he decided to call her and tell her himself that he wouldn't stand in her way, that he would make the divorce as easy as he could.

But there was nothing easy about the phone call. From the moment her voice echoed through the cold and heavy plastic hand piece of one of the shared pay phones in Thad's pod, he felt his chest seizing up. He no longer had the same feelings for her that they'd once shared, but hearing her voice, so bright and alive and normal, filled him with memories: of the apartment they'd shared, of the charity bike ride across the country, of nights spent in a tent, of their rushed wedding to escape his parents' anger, and most of all, those brief moments when she would warm his hands against her stomach, flesh against flesh.

But standing there, with the shouting and howling of the other caged animals all around him, the din of prison life echoing off the metal and cement, he couldn't say anything to her except that he was sorry, that he hoped she could be happy. And then, when it was her turn, hearing the noise of the prison behind him, she responded with the only words that she could think to say.

"Well, I hope you're having fun with your new friends."

And that was it; Thad was left standing there holding the dead phone in his hand. Sonya had no way of comprehending how terrible what she had just said had seemed to Thad, how completely alone and separated from the world it made him feel.

But he didn't have that long to dwell on the thought, because shortly after that phone call, he'd gotten word from his lawyer that the time had come.

The next morning, Thad would finally get his day in court.

40

The minute Thad saw the look in his lawyer's eyes, he knew that something had gone horribly wrong.

The day had started ridiculously early. At three A.M., they had come to put him back in handcuffs and shackles, to lead him back to the courthouse holding cell, a place he had been to a number of times over the past year and three months. The courthouse itself had become a place of total sensory overload to him, and by the time he was walked through the halls and into the courtroom, he was in a state of disassociation; his mind was so used to the dull environs of prison, he could barely comprehend all the colors around him, from the designs on the carpets to the clothes of all the people. It was really hard to concentrate on anything anyone was saying to him—and it wasn't until he looked at his lawyer, halfway through the proceeding, that he realized that terrible things were going on in front of him.

It wasn't just the monetary value of what he had stolen that was being talked about; after a parade of specialists in previous proceedings, the court had valued the moon samples somewhere between $7 million and $20 million, based not on street value, which would have put it much, much higher, but on the cost of the moon landing program, and the amount Thad had stolen, as a fraction of the overall mass of samples that had been brought back. So in that regard, Thad was pretty lucky;

it was a big heist, but it wasn't nearly the half a billion dollars' worth it could have been. But the value of the moon rocks and the Mars sample wasn't the issue that was terrifying his lawyer—it was what the judge was considering doing with the sentencing guidelines.

Though Thad could only piece a bit of it together at the time, because of his confused mental state, it turned out that the judge was considering giving a "5k2.7 enhancement" to his sentence, which, in layman's terms, was an additional rise in sentence due to a crime that "shut down a branch of the U.S. government." It was a rare enhancement that usually applied to terrorists—people who blew up federal buildings or killed important government employees.

Immediately, Thad's lawyer made an impassioned argument that such an enhancement was absurd, unfair, and illegal. But the judge was listening with deaf ears—and it wasn't the lunar rocks themselves, or even the Martian meteor, that was pushing her toward such a draconian judgment. It was those green notebooks that Everett Gibson had told the FBI about, the ones that Thad couldn't remember finding in the safe. Gibson had made an emotional speech at an earlier court proceeding—and the judge had decided that the loss of those notebooks, combined with the temporary loss and possibly permanent damage to the samples, was enough to warrant the charge.

"This is not an ordinary situation," the judge exclaimed, looking right at Thad. "The significant disruption of a government function by Mr. Roberts stealing these moon rocks—personally, Dr. Gibson's testimony was heart-wrenching. All the work that he had done that was just for naught because Mr. Roberts decided to steal not only the lunar samples, but also all of his scientific work that had been written in those notebooks—and these were national treasures that are priceless."

Thad's whole world started to melt as he heard the words, like a Dalí painting come to life. To have someone talk about him like this—it had simply never happened before, at least not since he'd been disowned by his family and the Mormon Church. He had always been the one with so much potential.

"To get the same thing back," the judge continued, "the government would have to go back twenty or thirty years in the space program. We're not going up to the moon to get rocks and samples every day. And in fact, Dr. Gibson can never go back and get his notes, and they can't use the rocks for the same educational and scientific uses that they had before because they're now worthless. I mean, Dr. Gibson was practically in tears on the stand because his—everything he had worked for was all for nothing."

Thad couldn't believe the venom in the judge's voice. And truthfully, up until that moment, he had never considered the pain Gibson would suffer from his theft—he still had trouble conceiving of it as anything but a victimless crime. He and the girls had taken full scientific precautions when handling the samples—as much as they could. As Thad saw it, even Gibson himself had basically referred to the rocks as trash.

But neither of those responses was going to make any headway with the judge, who had obviously already made up her mind. Before she handed down her sentence, Thad requested and received the opportunity to at least apologize. Hopefully, if he worded it just right, he could get the judge to be lenient. To show mercy.

"Sorry, I'm very nervous," he began, speaking as loud as he could. He hadn't strung that many words together in a long time, and his throat hurt with the effort. "But, Your Honor. I believe you have a very—of course, you've been presented all the bad things I've done in my life, but your image of me has been very shaded. It makes me very uncomfortable to even talk to you. But from what I've been hearing today, I think it would be important for you to know that the reason I even considered taking these moon rocks out of that cabinet was Everett Gibson had shown them to me a year before. He had, because of my enthusiasm, informed me that they'd been there for a long time, and he was basically in charge of just leaving them there, and he let me know they weren't being used."

He was warming up as he went, because it was the first time he had the attention of a crowd that he considered his peers—intelligent, educated people—since the JSC. It wasn't like being in a swimming pool full of co-ops, but it was something; it felt like, after fifteen long months, he was at least someone.

"And I'm not trying to justify my actions, they were definitely wrong. But I'm just trying to give you some kind of perspective of where I was coming from in there. I'm embarrassed and ashamed of my actions. I came into that whole thing, obviously, very naive. I've never had a chance yet to apologize."

And he was really off, now, into a monologue that had been building since he'd been arrested. In some ways, it had been building since he'd first set foot in NASA. Because he'd never felt like he'd belonged; he'd always felt like he needed to apologize for just being there. Hell, maybe the need to apologize went even further back, all the way to the beginning for him, all the way to Sonya and beyond that, to his parents, all the way back.

"I have somebody to apologize to. I'd like to quickly take that chance first to apologize to NASA for embarrassing them and for any trouble any individual had to go through because of my actions. And especially for abusing the trust that I had, between so many individuals there. So many people that were my mentors and my heroes are now very disappointed in me because of the potential they saw in me and encouraged in me. And at a weak moment, I did the wrong thing and abused that trust. And I still believe NASA is a wonderful organization. It inspires millions of people around the world to achieve higher goals and, you know, to reach for higher things. I still have a complete respect for them, and now I have to think of myself as a person who did this to basically my hero organization. And at the same time, I took away my own dream of being an astronaut."

He kept expecting the court to stop him, but this was his moment, probably his last; nobody was going to say a word until he was done.

"I think I should also apologize to science. At the time, I had tried hard to justify my actions, thinking because I knew these samples were already consumed—it doesn't justify the disgrace and embarrassment I brought to NASA and to science as a whole."

When he was done, he realized there were real tears burning at the corners of his eyes. But he could also tell that the judge was unmoved.

It wasn't until his appeal that he realized that no matter what he said, no matter what he'd ever say, his own explanations and apologies for what happened would never be able to stand up against what NASA felt he'd done; or, more specifically, in Everett Gibson's own words, given in a tearful victim's statement that entirely sealed Thad's fate:

"As an employee of the United States government, of NASA, and a research scientist, I would like to note that in 1969, some very brave individuals went to the moon and began recovering lunar samples—samples which are national treasures. They have been worked on in research projects, viewed by the public around the world with pride. It hurts me a lot to know that one individual would want to take it upon himself to steal one of these samples and benefit from it financially, knowing that this has hurt a large number of people. It has hurt our nation to know that we have one amongst us who's working as an intern in our own laboratories—that just broke our trust. I, as a scientist, have been hurt deeply. I, as an American citizen, am deeply moved and shaken by these actions which occurred. It hurts me deeply. Thank you."

That, more than anything, was what was going to enhance Thad's sentence beyond anything he could have expected. In the eyes of the court, in the eyes of Everett Gibson, he had committed a crime against the entire country—the entire world.

"It is the judgment of the court," the judge said, looking Thad right in the eyes as she lifted her gavel, "that the defendant, Thad Ryan Roberts, is hereby committed to the custody of the Bureau of Prisons for a term of one hundred months."

The gavel slammed down, and in that moment, with the explosive crack of wood against wood, Thad went completely deaf.

. . .

By the time he was led back toward his cell, Thad had a strange smile planted across his face—a mixture of disbelief, shock, and even a little relief, at finally knowing his fate, finally being able to give up what little hope he had left. As he made his way into the Submarine and down the hallway, the prisoners who could see him began shouting to one another, "Moon Rock, Moon Rock," because they knew their gambling game was about to be decided—and from the smile on his face, they all believed that he was about to give them a number that would hit at least some of their guesses. In fact, as he was led into his cell, some of his pod mates were already congratulating him, guessing from his expression that he was getting out with time served, fifteen months. When everyone had quieted down, Thad gave them the news.

"One hundred months."

There was laughter all around, because nobody believed him. They began peppering him with questions, demanding the real truth, but he didn't say another word, he simply crossed into the bedroom and lay down on his bunk.

It wasn't until the five o'clock news, when the prison population learned he had been telling the truth—indeed, Moon Rock had gotten a sentence of more than eight years in federal prison—that they all grudgingly got down on the floor and began doing push-ups.

41

A few days after his sentencing, Thad received two pieces of news that, together, were just enough to keep him from contemplating suicide. First, he was being transferred out of the Submarine. Because he was now a sentenced federal prisoner, he was going to be taken to a midlevel security camp, which had to be better than the county jail where he had been held for the past fifteen months. But that news paled in comparison to the news his lawyer gave him at their next meeting.

Rebecca had received permission from her probation officer to talk to him one last time.

Thad memorized the phone number his lawyer gave him, intending to make the call as soon as he got back to his cell. But by the time he'd been uncuffed and unshackled, he'd missed his chance at the pay phones; he was forced to spend the next eight hours of sack time—the last hours he'd spend at Orient—sleepless and tossing and turning against the steel bunk.

The transfer to the federal penitentiary went by in a blur. Thad briefly remembered being on a Continental flight, chained up next to a frightening-looking man who just wanted to hear stories about moon rocks—and then he was being led into his new home, where he'd be spending the next phase of his life. And it was true, the federal camp was much better than where he'd come from; there were two-to-four men to a cell, and there were multiple television rooms, well-kept outdoor areas, and best of all, porcelain toilets—with real seats.

But the real difference between county jail and the federal prison was something Thad discovered a mere hour after he'd been checked into his new cell. Although it was the designated lunch hour, he had chosen to stay behind to take care of a bit of business he hadn't been able to get to during his flight over from Orient. He was seated on the porcelain toilet, halfway into what he needed to do, simply reveling in the idea that there was no one standing a few feet away, playing cards, cracking jokes—when a guard suddenly stuck his head into the room. A tray had apparently gone missing from the lunch area, and since just like in county, all trays had to be accounted for—the guard had been sent to check the cells. But seeing Thad seated on the toilet, the guard did something that took him completely by surprise.

He gave Thad an embarrassed look, and turned away.

"Sorry, man, I'll come back when you're done."

Thad sat on the toilet in shock. It was the first time he'd been treated like a human being in more than a year.

. . .

Twenty minutes later, he was in front of another pay phone—this time a phone that was situated in a cubby carved into one of the cinderblock walls, separated from a TV room by a low paneled divider. It was a level of privacy that Thad hadn't enjoyed for quite some time, but it didn't make him any less nervous as he dialed the number.

He had practiced what he was going to say, but he was pretty certain that as soon as he heard her voice, he was going to forget everything he had planned.

He wanted to tell her that he expected her to move on; he wanted her to go and live a happy life. He knew, now, that he was going to be gone for a very long time. He wanted to tell her that he loved her, but that he would understand, she was young, she needed more. It was going to be a hard conversation, but it would also help him find a way to deal with what had happened, where he now found himself.

Since it was a collect call, as soon as he finished dialing, a mechanized voice came over the line—indicating that the person receiving the call had to hit the number five in order to accept the call, or seven to refuse. At the proper moment, Thad spoke his name for the recording, then listened as the operator put the call through. Two rings, and Rebecca picked up, but before Thad could say anything, she hit a button—and the phone went dead.

Thad felt like he'd been kicked in the gut. She'd hit seven. It didn't make sense. His lawyer had told him that Rebecca had wanted to have the phone call. Thad quickly redialed the number. He went through the same routine, giving his name; this time Rebecca picked up on the first ring. And this time she hit the right button, because her voice suddenly splashed in his ear.

"I'm so sorry. I thought I was supposed to hit seven. I heard it wrong."

In that instant, as he had suspected, Thad forgot everything he wanted to say. She sounded so close, like she was standing just a few feet away, and her voice brought him spiraling back that year and a half, even further, to their first date, to an image of her pointing out fish in an aquarium, to her smiling reflection against a thick wall of glass.

They talked quickly. He told her that he still loved her, and she responded that she still loved him. He told her that she was free to do whatever she needed to—and she responded that she didn't want to think about any of that, that all she could think about was him.

As the collect-call limit drew nearer, Thad rushed to say the thing that was most important to him.

"I need a way to communicate with you. There has to be some way. And if it can't be you directly, if it has to be a friend that I'm talking to, that's fine. There just needs to be some way. I need that to survive in here."

"But my father—"

"Rebecca, there has to be some way."

Rebecca finally relented; she gave Thad her sister's address, speaking slowly enough so that he could memorize it as she went.

"I'm going to write you every day," Thad whispered. "And the letters will be my lifeline."

Before Rebecca could respond, before Thad could tell her one more time that he loved her, the operator's voice cut in—and then the line went dead.

Beautiful Rebecca,

I hope you find yourself living a dream. I think of you often and send my love out into the unknown, hoping that somehow it finds you and warms you with a smile. I hope you have not let trouble convince you of impossibilities. There is no dream beyond your grasp, Rebecca. You are the rarest type of person there is and you deserve the best that emotion and experience can offer. Someday I hope to learn that every day finds you laughing, that your path matches your dreams, and that you have discovered that your fate isn't to be an old lady with a few cats, but to live in passion to receive love, companionship, trust, and comfort to the degree that those fires live in you . . . the ones I knew briefly. Although it pains me to imagine you with another, it hurts more to imagine you living without love.

42

And for the next year, it was those letters that kept Thad sane. Through the flowery, sometimes clichéd, but always sincere missives, which he toiled over for days on end—writing and then scratching out words, phrases, sometimes entire pages—he was able to hold on to his sense of self way past what either he himself or jailhouse wisdom could have predicted. Those letters really were lifelines, even if they were entirely one-way. Thad was able to artificially keep alive the character he had created at NASA, the romantic, adventurous, fantasy persona that would never normally have been able to exist in a place like prison. He was surrounded by animals, but when he finally found a moment alone, holed up in his bunk or in a corner of the laundry room, or even on the toilet, he could go back to that place and become the person whom Rebecca had fallen in love with.

He never got a response, not a letter or a message via his lawyer or any sort of phone call. But the letters he wrote were enough, because they allowed him back into that place where he was most powerful, his own mind.

It was because of that inner strength that he was able to embark on what he would later see as a revolutionary journey—which began, really, as just an attempt at finding a way to keep busy in between letter-writing sessions. Leafing through the adult education manual that was given out to all inmates who had been in the federal system long enough to qualify for classroom privileges, he quickly realized that there wasn't anything advanced enough for someone with his background. So instead of taking a class he was overqualified for, he decided that maybe there would be a way for him to share his own knowledge with others.

He lobbied the warden and the heads of the adult education program, and he eventually received permission to teach an astronomy class—the first of its kind in the federal penitentiary—to any inmate who was interested in the stars.

The first day of class, Thad arrived at the small, windowless classroom not knowing what to expect. To his surprise, he found the place crowded; his notoriety as the guy behind the Moon Rock Heist had appealed to inmates who wanted to hear stories about NASA, spaceships, and often alien life. From the very beginning, Thad used the inmates' eclectic interests to guide them into a more basic study of space and the unknown. Because they couldn't exactly go outside at night to look through telescopes, he focused on the many theories behind the science of astronomy, and did his best to get the inmates excited about the mysteries of the universe—things like black holes, supernovas, and dark matter.

He had only one requirement from his students. If they enjoyed his class, when they eventually got out of prison, they were each to send Thad one physics book—so he could continue studying the subject he had found the most challenging of his three college majors. Since, as a prisoner, he could only keep up to five books in his cell at one time, he had other inmates hold them for him, rotating through as many books as he could read, as quickly as he could get them.

Week after week, month after month, he taught astronomy and spent his nights reading physics—and slowly he found himself focusing on the current state of quantum theory. It was a topic he had been introduced to back at Utah, before he'd distracted himself with other pursuits; given an almost infinite amount of time, and a pretty good collection of the current literature, he set out to devise his own new theory, to make better sense of the things that he found missing from the accepted liturgy.

Some men found God in prison, others found themselves—but Thad threw himself into advanced physics, which led him to look at the world in a new way. He was intrigued by the fact that when physicists studied very small things—quanta the size of atoms—these objects were characterized by a certain level of indeterminacy. Stimulated by further readings on quantum mechanics, Thad began, in the simplest terms, to look at the world of these tiny particles—from their perspective.

From a distance, the image of a Teletubby on a TV screen appeared continuous and fluid; the closer you got to the screen, the easier it was to see that in fact, the image was made up of tiny pixels—but still, the pixels seemed part of a continuous whole, connected to one another on every side. But when you got even closer, so close that *you* were the size of one of those pixels—you realized that in fact, the pixels were not set into a static plane, or part of a continuous whole; they were individual units adrift in a sea of similarly tiny quanta. To describe these individual units correctly, and stirred by his readings on string theory, Thad began to learn that you needed to throw out the idea of four dimensions, and move to a more accurate theory involving eleven—nine of space, and two of time—and even formulated some ideas of his own.

Thad's prison astronomy students did not have the physics background to begin to understand a multidimensional way of looking at life, but the classroom sessions still became a passion for him, because it was a place where he could go to work out his ideas, and to inspire people to at least begin to fantasize about a world beyond the prison walls.

As the months passed, Thad settled into his new routine, teaching, writing, and always reading—and despite where he was, despite his sentence, he began to carve out a life that he could tolerate. And he continued like that, complacent if not content—until the day that one of his cell mates approached him at the end of astronomy class to tell Thad that he'd received a sizable allotment of mail. Even so, Thad expected nothing more than a package filled with physics books, sent from an overly grateful ex-student.

But as soon as he reached his cell door—he saw that it wasn't books at all.

To his utter shock, there, on his bunk, stacked together in a pile more than a foot and a half high, were all of the letters he had written to Rebecca. Posted but unopened, every one of them marked *return to sender, address no longer valid*.

Thad stood there in the doorway to his cell, unable to breathe. Rebecca hadn't read any of them. Either her sister had moved and left no forwarding address, or she had simply refused to send them along to Rebecca. Thad had been writing into a vacuum, pouring all his love and passion into nothing more than a cosmic black hole. Rebecca was gone, and he would probably never hear from her again.

And in that moment, the last connection to who he was before vanished, the last strings tying him to his old life severed, the persona he had built up through equal mixtures of hard work and fantasy emptied out of him, and he collapsed to the floor of his cell.

I will always cherish the experience of you. I will always love you and wish for your happiness—even when I cannot be a part of it.

Please allow me some closure, Rebecca. It's time for this wound to heal.

Wishing you wide-eyed wonders, love, and contentment . . .

Love,

Thad Roberts

43

Axel had just walked in from the popinjay field, his shoes caked in grime and his thick, meaty shoulders aching from the crossbow, when he saw the little package on his front stoop. He knew before he even saw the address whom it was from, because the markings all over the manila packaging were as easy to recognize as a René Magritte. It was from overseas—which meant America, because the only people he knew overseas were in America. And since there were no official seals imprinted anywhere on the thing, he knew it wasn't from the FBI. But it *was* from a government agency.

Dr. Everett Gibson had first reached out to Axel right after Thad Roberts had been sentenced to more than seven years in federal prison. At first, Axel had harbored mixed feelings when he'd read about the harshness of Thad's sentence; after all, the kid hadn't really been the master criminal Axel had pictured, he'd been naive and foolish, maybe a bit arrogant, and certainly misguided. He hadn't physically harmed anyone, and the samples had been recovered.

But the crime he had committed—it wasn't like stealing a car; it had involved a national treasure. Taking those moon rocks was like slapping his country across the face. And after meeting Dr. Gibson in person—as a reward, the esteemed scientist had actually come over to Belgium and spoken to Axel's mineral club about the ALH meteorite and the possibility of life on Mars; boy, the youth center had been busting at the seams that snowy night!—Axel had finally decided that maybe Orb Robinson had gotten what he'd deserved.

Everett Gibson had suffered greatly because of the Moon Rock Heist; at the time it had gone down, he had been in Australia on vacation, and upon landing back in the United States, he had been taken by the elbow on both sides by federal agents, interrogated, and wholly embarrassed by what had occurred in his lab. Apparently, there had been a series of numbers affixed to the top of his safe, which Thad had wrongly suspected to be the combination. In truth, they were a simple algorithm: all you needed to do was take the square root of the numbers and triple them, and you had the combination. But just seeing those numbers may very well have inspired Roberts to think he could succeed in the crime.

And Gibson had lost more than face; the night of his lecture at the mineral club, he'd nearly had tears in his eyes as he told Axel about the missing green notebooks that he still, to this day, believes Thad Roberts destroyed. At trial, Roberts had denied ever seeing those notebooks, and Axel would never know for certain what the real story was. But Gibson was a respected man of science, and Axel took him at his word.

At the "Mars in Antwerp" lecture, Gibson had presented Axel with an official plaque thanking him for, essentially, saving NASA's bacon; and along with that, a framed photo of a lunar landing, signed by a real astronaut himself! And to Axel, that would have been enough.

But standing on his front stoop, tearing into the manila package with his blistered archer's fingers, he quickly discovered that Gibson had one more little symbol of his gratitude to bestow.

Inside the package was an official letter, stating that Dr. Everett Gibson's request to the International Astronomical Union had been approved. They had renamed Asteroid 15513—which would now orbit the sun under the name "Emmermann."

You will live forever in the heavens between Jupiter and Mars, Dr. Gibson wrote.

It was an incredible thing. The very idea—unimaginable!

There was a rock between Mars and Jupiter that was named after Axel. Seven kilometers long, two kilometers wide. Axel would never see it, or touch it, or visit it, but it was there, and it would always be there. Spinning through the vast emptiness of space, forever.

EPILOGUE

Deep into a seven-and-a-half-year sentence, the only dimension that really mattered was time, and it wasn't measured in minutes, hours, days, or even years, it was measured in seasons—because the seasons were something you didn't need to mark on a calendar or scratch into a cinderblock wall. The seasons you felt against your skin and in your bones, during the brief minutes you got to spend outdoors, milling about a rec yard or playing cards at a picnic table, and also late at night, listening to the wind or the rain or even the snow whipping endlessly against the steel-and-concrete exterior of the prison walls. The seasons were something real and unavoidable, and they couldn't be controlled by a hack in a uniform or a judge in flowing robes.

At Florence Federal Prison in Colorado, located just ninety miles from Denver, the season that had the most resonance to Thad was winter; snow so deep you could wade through it, the air brisk enough to wake you from the numb monotony of life in a cage. And although this medium-security compound didn't have walls and catwalks and guard towers or even fences around its perimeter, it was still a cage, one of a half dozen Thad had been transferred to and through over the past few years of his life since NASA.

Overall, life in Florence was as tolerable as Thad had experienced since his disintegration and slow, internal rebuilding after finding his letters to Rebecca returned, unread and unopened. He had survived that moment, somehow, but it had taken months before he'd resumed his teaching, reading, and contemplating—not as the person he was at NASA or before, but a numbed, yet stronger version of himself.

It was in the midst of this reformation, in the middle of a winter that seemed to go on forever, that Thad also found himself reconnecting with the outside world, in the form of an acquaintance from his school days at Utah, a bright, adventurous kid named Matt who had shared a few physics classes with him back before he'd even gotten the job as a co-op at NASA, one of the few people—if not the only person—who had not completely forgotten about Thad. For whatever initial reason—curiosity, sympathy, genuine kindness—Matt had reached out to Thad in prison, first in the form of letters, then in fairly frequent visits, building what had become a true friendship, or as true a friendship as two people could have, separated by the federal justice system. Matt had remembered Thad as the brilliant kid in physics classes who was willing to go further and think freer than anyone else; and in the letters and visits, as Thad told him about his new multidimensional physics theories and his teaching in prison, Matt became intrigued by what Thad was doing, how he was again reinventing himself in such a dark, difficult place.

With time shaved off on appeal and for good behavior, the end of Thad's sentence was approaching, but it was almost impossible for him to think about life after prison in any real, concrete terms, but Matt had made it his mission to help him regain at least some of what he had lost. Still affiliated with the University of Utah, Matt set his sights on getting Thad back into the university so he could finish his undergraduate degree and then go on to chase a Ph.D.

To Thad's surprise, the physics department at Utah—especially the chairman of the department, a man Thad had known well and impressed when he was a college student—was initially very supportive of the idea. But there was a clear roadblock—the geology department fiercely opposed the idea of letting Thad back into school. During the trial, the fact that Thad had stolen fossils from the university museum had been part of the prosecution's arsenal against him—and the geology department had branded him a thief right along with NASA. Matt himself had been to a few of Thad's dinner parties, where Thad had shown off the fossils he had taken from the museum—Matt hadn't known at the time they were stolen, just that it was an incredibly impressive collection for a fellow student to have—and it was understandable that the geologists at Utah wouldn't want Thad back in their lives. But Matt also knew that Thad was a different person—that he had served his time, that he had been punished far beyond what Matt felt he had deserved.

A handful of professors at the university agreed, especially in the physics and philosophy departments—but still, it seemed impossible; Matt simply couldn't get Thad reenrolled in the university. Not because he was the kid who had stolen the moon; even several years later, the geology department could not forgive Thad for the earthly rocks he had taken from the bowels of the museum.

Still, with Matt's help, Thad made the idea of reenrolling in the university his new goal. He came up with a simple plan; once he was out of prison and placed in a halfway house for the few months of supervised release that would begin his probationary period, he would get a job on the university campus—anything, really, as menial as it had to be. He would offer himself up as a teaching assistant to the professors who still believed in him, the ones in the physics and philosophy departments who still felt he had the potential to do something important with his life. Eventually, they would see that he was serious, and they would let him reenroll. Not in geology, of course—he doubted they'd ever let him anywhere near that department again. But physics, philosophy, and eventually the philosophy of science, which was where he now wanted to go. He had always been a good student in the past, and he would one day prove that he could be a good student again.

August 4, 2008

A brilliant Colorado morning, the clouds like twists of cotton, the sun breaking through in beams so bright they played across the prison compound like strips of lightning.

It was a Monday, and the procedure started at ten, but Thad didn't actually let himself believe he was getting released, that it was really, finally happening, until he was truly on his way out of the prison—more than three hours later. He'd heard too many stories about other inmates who thought they were on their way to freedom, when something happened to gum the works, some sort of prosecutorial appeal. Even after so many years, Thad couldn't let himself believe that it was finally over. He had served his time.

Dressed in his greens—green pants, green button-down shirt over a white tee—and his steel-toed prison boots, he was led past the track that circumnavigated the prison yard, his final steps across the compound. He couldn't even begin to imagine how many times he'd run around that quarter-mile strip of dirt—just doing the math in his head, he knew he'd circled it so many times he could have run from L.A. to New York a dozen times. And then he was past the yard, being shuffled into a waiting van for the short drive to the processing unit. All he had with him, other than his prison greens, was his single belonging—his physics theory, compiled in a book that was now almost four hundred pages long, loose sheets of paper held together by a pair of rubber bands. He had the book tucked tightly under his arm as he was led through processing. He had no idea what he was going to do with it—but to him, it was more valuable than a safe full of moon rocks. He truly believed it was his future, his reinvention, his new self.

Once the processing paperwork was finished, it came time to get paid. Most inmates spent the money they made working in the prison—the twelve cents an hour they were paid to do laundry, bang out license plates, shovel rocks and snow. But during his years in prison Thad hadn't needed anything other than books, which he hadn't been allowed to buy—so he'd saved up more than a thousand dollars in his prison account. He felt a little burst of excitement as he watched a young officer behind a desk count out the money from a register—until he saw the bills themselves.

"What the hell—is that monopoly money?"

The officer laughed, shaking his head, explaining that in the years Thad had spent locked up, the government had changed the look of fives, tens, and twenties. Thad realized with a start that he hadn't even seen a single dollar bill since he went to prison. No doubt that would just be the beginning of his culture shock; he'd been in a time capsule, a state of stasis—the world wasn't going to look the same as it had when he'd been sent away. It was a terrifying, sobering thought.

After the processing unit, he was led back into the van—and then it was finally, truly happening; he was leaving the compound for the short trip to the station where he would wait for the bus that would take him the first part of his journey back to Utah. He spent most of the van ride simply staring out the window, watching the prison compound until it had receded into the horizon, as it went from three dimensions to two, to one—just a pinpoint at the farthest reaches of his vision, nothing, a memory.

An hour later, he wasn't a prisoner anymore, he was just a guy sitting on a bench waiting for a bus. But he didn't wait long—even though the bus wouldn't arrive for another couple of hours, he couldn't sit still after so many years in prison; he couldn't spend another moment in frozen isolation. It was against the rules—already, just after his release, he was technically breaking the law—but he'd arranged to have one of his astronomy students and closest prison friends who had been released a year earlier, a former gangbanger named Joey, pick him up at the station. He hadn't planned on really going anywhere—but Joey had taken care of the details for him. A few miles from the bus station was an Olive

Garden restaurant—they'd have no problem making it there, having lunch, and making it back in time to catch the bus.

It was the most fascinating lunch of Thad's life; the food, the people, the noise, the colors—even the walls, so different from the white on white he had grown used to—it was all one massive, distracting, mind-blowing sensory overload. He didn't even know what he was eating, just that there was so much flavor and heat, and it kept on coming, until he could barely stand up from the table and follow Joey back out to the car. Everything felt so surreal. Even as he shook Joey's hand, thanking him for his first real, truly free moment in aeons, he felt like he was in some sort of dream, that any moment he'd wake up in his bunk back in the prison, staring at the white-on-white walls.

But instead, he went from the Olive Garden to the back of a bus heading to the nearest airport, his physics manuscript still tucked under his right arm. His body was sated by the heavy meal, but his mind was still racing. He had no idea what was next, but the world seemed so open, so new.

He felt the weight of the physics manuscript against his arm. He knew that there were people who would say it was nothing but another one of his fantasies, another game of his mind finding its way into reality. A dream, even a con—yet another reinvention.

A fantasy—like the idea that a kid from nowhere, from nothing, could somehow believe that he could one day be an astronaut, that he could one day be the first man to walk on Mars.

That this brilliant, enthusiastic, impetuous kid could fall so deeply and fully in love with a girl he'd only known for a month—that he'd be willing to throw it all away.

A fantasy, a dream—maybe even as impossible as stealing a piece of the moon.

ACKNOWLEDGMENTS

First and foremost, I am grateful to Thad Roberts for opening up his life to me over the many months it took to research the facts of this amazing story; it simply could not have been told without his generosity and his honesty. Likewise, I am indebted to Matt Emmi, for guiding me into Thad's world, and to Bill Flagg, Eric James, and of course Niel Robertson—from Vegas to Silicon Valley to NASA, Niel really is my ace in the hole.

I am indebted to Bill Thomas, the most amazing editor one could ever ask for, and the team at Doubleday/Anchor, especially Melissa Danaczko, Todd Doughty, and Russell Perrault. I am also incredibly grateful to Eric Simonoff and Matt Snyder, the best agents in the business. Again, many thanks to my Hollywood brother Dana Brunetti and the brilliant Kevin Spacey, as well as Scott Rudin, Mike De Luca, Amy Pascal, Doug Belgrad, Elizabeth Cantillon, and all the wonderful folks at Sony. I would also like to thank my secret weapon—Jeff Glassman—and his associate Michael D'Isola. Many thanks to Barry Rosenberg, Megan Cassidy, and to my brothers and their families.

Furthermore, this book could not have been written without the generous help of numerous inside sources, especially the immensely affable Axel Emmermann, Gordon McWhorter, and many others who have asked to remain anonymous. I have always been fascinated by

NASA, and I did my best to capture the wonder, beauty, and genius of an institution that I respect and admire. I hope we, as a nation, continue to generously support this amazing group of scientists; we have all greatly benefited from each step forward, be it the first moon landing, the shuttle program, or the quest for Mars.

As always, I am grateful to my incredible parents, and to Tonya, Asher, and of course, Bugsy—simply put, you guys make it all worthwhile.

ABOUT THE AUTHOR

Ben Mezrich is the *New York Times* bestselling author of *The Accidental Billionaires* and *Bringing Down the House,* in addition to nine other books. The major motion picture *21,* starring Kevin Spacey, was based on *Bringing Down the House.* The Oscar-winning film *The Social Network* was adapted from *The Accidental Billionaires.* Mezrich lives in Boston with his wife and son.